Mm

(Much Ado about Nothing)

A. G. Davis Philip
Union College and Van Vleck Observatory

Adam Robucci
Union College

Michael Frame
Union College

Kenelm W. Philip
University of Alaska, Fairbanks

Series on Fractals - Volume 1

Midgets on the Spike

L. Davis Press
Schenectady, New York

with 132 figures

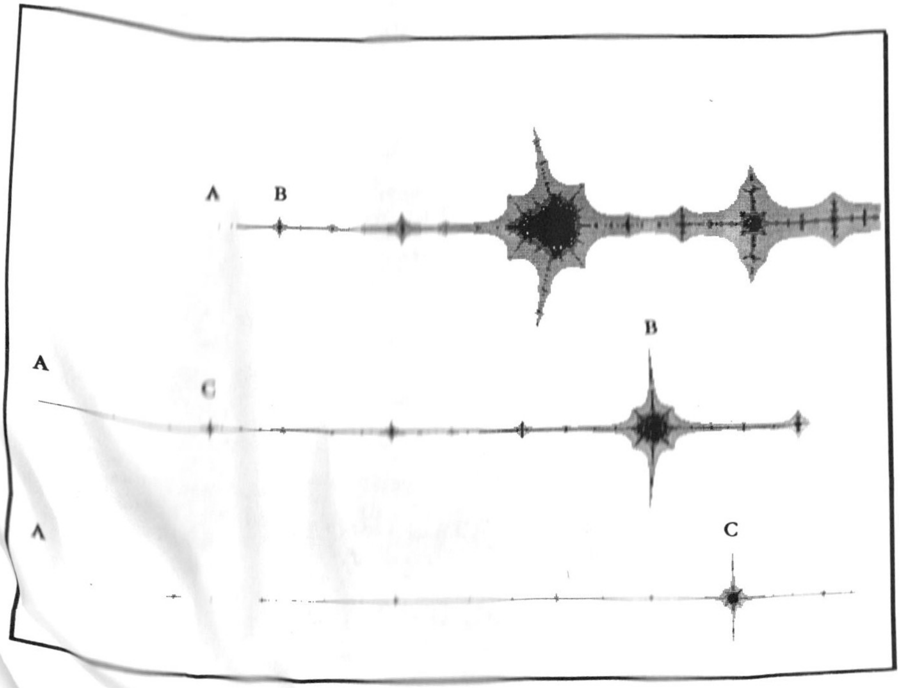
A
B
A
C
B
A
C

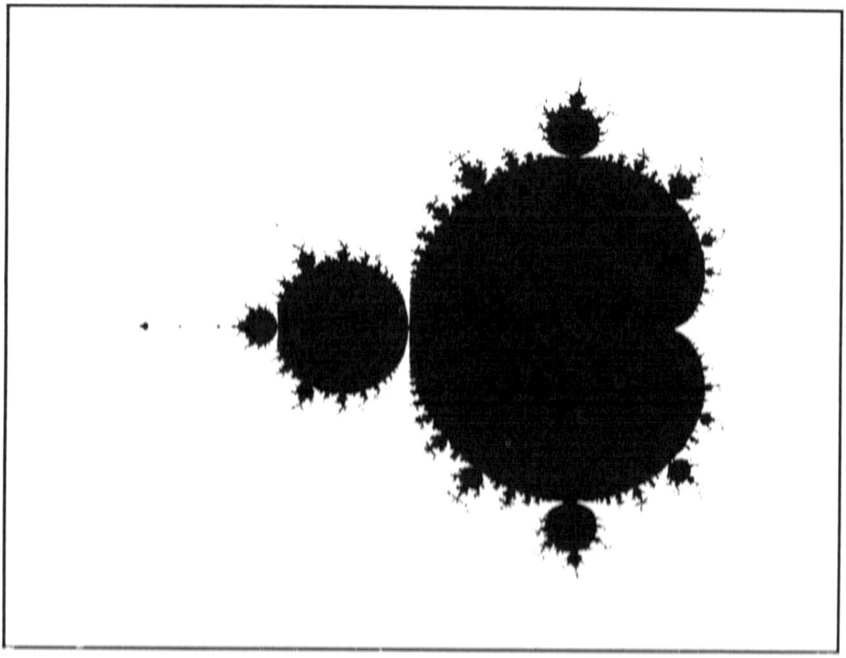

L. Davis Press

Library of Congress Catalog Card Number 91-092943

TABLE OF CONTENTS

PREFACE

In 1985 (Dewdney 1985) in the Computer Recreations section of Scientific American there was an article introducing readers to the Mandelbrot set. This article was the incentive for many investigators who started their own searches of the set. In 1988 D. Philip (AGDP) and K. Philip (KWP) each obtained computers that were capable of computing EGA and VGA resolution images, to magnifications of about 10^{16}. We each started on a program of investigating the major features of the Mandelbrot set and we have reported on many aspects of our work in articles published in Amygdala, the newsletter published by Rollo Silver in San Cristobal, New Mexico. We have also gone on three joint lecture tours (Alaska, East Coast and West Coast), giving a lecture entitled, "An Introduction to the Mandelbrot Set". One of our investigations concerned the Midgets found on the Spike of the set and in this volume we describe the results of our studies as the magnifications were pushed to ever higher numbers.

At Union College, Michael Frame, of the Mathematics Department, teaches a course involving the study of chaos and related subjects. AGDP, for the last three years, has shown some of the tour slides to these classes. We were able to interest some students to do their class research projects on a subject related to some aspect of the study of the Mandelbrot set. In 1990 Adam Robucci joined the team and he was responsible for writing the high precision routines that enabled us to extend the midget search to magnifications as high as 10^{359}. He also wrote other programs that have been used in the analysis and display of the images.

We had now assembled a team of two empirical investigators, one mathematician and a programmer and this combination of talents has enabled us to make the investigations reported on here. The electronic network, BITNET, has been used to keep the Schenectady members in communication with KWP in Alaska. Two papers related to this project are in press or have been published [Frame et al. 1991 and Philip and Robucci 1991] and readers are referred to them for further details.

Many of the most familiar fractals had their origins in the pure mathematics of the turn of the century. Cantor constructed the Cantor middle-thirds set in 1883 to show the line contains sets consisting of uncountably infinitely many points and yet having no length. To warn that the notion of dimension needed a careful treatment, in 1890 Peano showed there are continuous curves filling up the entire plane. In 1904 von Koch developed the Koch snowflake to prove there are continuous curves with no tangent lines. The Sierpinski gasket of 1915 arose as another type of curve with peculiar properties. Although these provided examples for Hausdorff's 1918 foundational work on sets with noninteger dimension, in general they were viewed by mathematicians as pathologies, as monsters to be avoided. Sets like these arose naturally in the work of Fatou and Julia (1910s and 1920s) on iteration of complex functions, but lacking any computer facilities they were unable to make concrete whatever mental images they had The hypnotic visual appeal of fractals had to await someone acquainted with the relevant mathematics and with access to good programmers. Awareness of fractals' potential as powerful tools in science had to await someone familiar with a wide enough range of science to perceive the presence of self-similarity in a vast array of physical phenomena. Perhaps it is surprising (or perhaps it was inevitable) that these are the same person. That person is Benoit Mandelbrot.

How the Mandelbrot set came to be discovered has been recorded in several places (Peitgen and Richter 1986, Peitgen and Saupe 1988) in Mandelbrot's own words. Here we give a brief sketch, striving to communicate some of the wonder involved in the emergence of the first pictures of the Mandelbrot set. In 1978 Mandelbrot and his assistants began computer experiments to draw solutions to iterative geometrical problems described by Poincare and by Julia. After this, the first experiments in parameter space problems involved studying for which c the function $c(z^2 + 1)/((z(z^2 - 1))$ has an attracting cycle. Very interesting pictures emerged, and Mandelbrot tried the same problem for the simpler function $z^2 - c$. To speed up computations, this was replaced by the related problem of finding those c for which the orbit of $z = 0$ does not diverge to infinity, first implemented for the function $cz(1 - z)$. For this function, some components of the set could be determined analytically, and the early computer-generated pictures agreed with these results. With each additional computation, a wonderful array of hierarchical features appeared. Eventually, though, with increasing magnification the pictures became fuzzier. Run on faster machines at IBM, what had at first seemed to be problems resolved into smaller copies of the whole set. Redoing the calculations for $z^2 - c$ revealed the now familiar pattern of midgets (though left-right reversed from current usage: now $z^2 + c$ is the function iterated). Many other wonderful things appeared, like evidence that the set is connected (proved by Hubbard and Douady) and that the boundary of the set exhibits a "hieroglyphical" dictionary of many corresponding Julia sets (proved by Tan-Lei). Still, though, imagine those amazing days, when a whole universe of infinite diversity opened before Mandelbrot's eyes. This volume presents some of our investigations into one set of features of this universe.

The manuscript was prepared on an IBM Model 80 PC using WordPerfect 5.1. The images have been printed on laser printers (LaserWriter and HP LaserJet). The color picture was photographed from a 3D image on the screen of a Mac II. The 3D program was written in True Basic by KWP.

We wish to thank the following people who read the manuscript and suggested modifications: Michael Freeman (N. Vancouver, Canada), Scott Huddleston (Beaverton, Oregon) and Kristina Philip

1. INTRODUCTION

A picture of the Mandelbrot set **(M)** is shown in the center panel of Fig. 1, along with four small pictures, one in each of the corners, which show examples of some of the interesting areas that can be found in the set. The shape shown in the central panel is generated in the complex plane by iterating the equation $Z \rightarrow Z^2 + C$, where Z and C are complex numbers. The starting value for Z is $0 + 0i$ so the first iteration gives for Z the value of the constant C. The second iteration yields $C^2 + C$, the third $(C^2 + C)^2 + C$, and so on as more complicated expressions of the complex numbers are formed. What is interesting in all this is what happens to the successive values of Z that are calculated. For some choices of C the newly iterated values of Z move rapidly away from the origin and approach infinity. For some other choices of C the values of Z converge to cyclic behavior, for others the values wander seemingly chaotically through a bounded region of the plane. The points C for which the values of Z do not diverge to infinity are called members of **(M)**. These are the areas marked in black in the center panel of Fig. 1. Areas colored white are occupied by pixels which are non-members and iterations started at these points rapidly move away from the black areas. An article by K. W. Philip (1988) (concerning a program called WORMTRACKS)) describes the tracks made by successive values of Z as the equation is iterated (see Fig. 4 and discussion). By studying the patterns made by these tracks many interesting things can be learned about the location chosen for investigation in the set.

The Mandelbrot set was discovered by Benoit Mandelbrot, who introduced the new mathematical field of study concerning fractals. Readers interested in learning more about the history of this subject are referred to the following books [**Chaos** by Gleick (1987), **The Beauty of Fractals** (Peitgen and Richter (1986), **The Science of Fractal Images** by Peitgen and Saupe (1988), **The Fractal Geometry of Nature** by Mandelbrot (1983), **Does God Play Dice?** by Stewart (1989), **The Emperor's New Mind** by Penrose (1989) and **The Problems of Mathematics**, by Stewart (1987)]. This monograph deals with one aspect of **(M)**, the midgets (tiny, but not exact, replicas of **(M)**) that appear on the Spike.

The parts of **(M)** are labeled in Fig. 1. To the far left is the Spike, on which can be seen the Biggest Midget. Going to the right one encounters an infinite sequence of heads (Head' and Head are labeled), an increasing cascade of tangent disks, terminating with the largest disk, attached to the large cardioid shape. The next feature is called Seahorse Valley, since it contains spiral shapes resembling seahorse tails. The largest part of the set is the Body, a cardioid shape to which are attached circular appendages pointing outward from the center. Mandelbrot has named these features atoms. Outward from the center of **(M)** there will be found other features attached to the atoms. The whole feature, atom and other structure, is called a radical. Each atom has a smaller set of radicals attached to it, and if these smaller radicals are enlarged it will be seen that their atoms have still smaller radicals attached.

The sequence of heads attached to the cardioid is an example of the same process mentioned above. Since this goes on forever as higher and higher magnifications are made, it exemplifies the fractal nature of **(M)**. At the far right of the cardioid is East Valley

Seahorse Valley

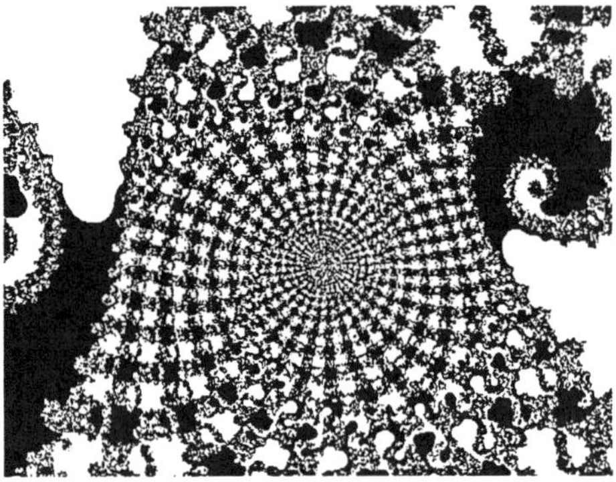

Seahorse Valley

Fig. 1. The Mandelbrot set and four selected regions.

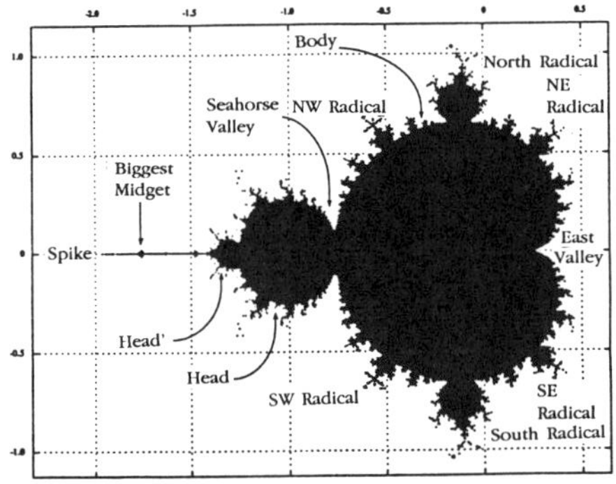

The Mandelbrot Set

Midget on a Spike

Seahorse Valley (Pseudo 3D)

(or Elephant Valley because of the elephant-like shapes found there). In Fig. 1. the four enlargements of various points in the Mandelbrot set (shown in the four corners of the figure) are taken from Seahorse Valley, from a tendril off a midget on the Spike and a pseudo-3D image from Seahorse Valley. They serve as an introduction to some of the diverse shapes that can be found in (**M**). As an aid to keeping track of the positions one can investigate in (**M**) Philip and Philip (1990) presented a scheme for naming the atoms which are found along the periphery of (**M**).

Fig. 2 (on the next two pages) shows the bifurcation diagram for $x^2 + c$, x and c real (lower diagram), together with (**M**)(upper diagram) where $z^2 + c$, z and c are complex. In both pictures, the c values are on the same scale. To obtain the bifurcation diagram, for each value of c the orbit (x_0, x_1, ..., x_{300}) of $x_0 = 0$ under iteration of $x_{n+1} \leftarrow x_n^2 + c$ is generated and the points (x_{100}, c), ..., (x_{300}, c) are plotted. The dotted lines indicate that the beginning of the bifurcation graph corresponds to the cusp of the main cardioid of the Mandelbrot set, that the first period-doubling bifurcation corresponds to the head-to-cardioid attachment, and that the second period-doubling corresponds to the head'-to-head attachment. Of course, there are many other such similarities between these two pictures.

The insert shows the analogous correspondence for the central branch of the Cycle 3 window of the bifurcation diagram and the Cycle 3 midget along the Spike of the Mandelbrot set. A cycle-n midget is one for which the iteration pattern for any point c in the midget cardioid converges to n points. We say n is the cycle number of this point c. Thus the Cycle 3 midget is one for which the iteration pattern converges to cycle between three points. The Cycle 3 Spike midget is the biggest midget on the Spike. These relations are not coincidences, but in both cases reflect the behavior of the orbit of 0 (the critical point for both functions). For example, for c taken along the single branch of the bifurcation diagram, the orbit of 0 converges to a fixed point. This corresponds to the observation that for c taken from main cardioid of the Mandelbrot set, the orbit of 0 converges to a fixed point.

2. MIDGETS ON THE SPIKE

Fig. 3 shows an enlarged portion of the end of the Spike, between real values of -2.00 and -0.9. The Cycle 3 and Cycle 4 midgets are indicated in the picture. One can find the cycle number of a point in (**M**) by studying the iteration tracks for that point in the region, as shown in the program WORMTRACKS (mentioned earlier). If the cycle number is one, then the iteration track pattern (formed by plotting successive values of Z as they are computed from the iteration of $Z \rightarrow Z^2 + C$) will converge to a single point. If the cycle number is two, then there will be two separate patterns, and so on. Another way to determine the cycle number is to count the number of rays in the star-like pattern above the head of the atom. For example, the North Atom has a pattern with three arms, thus its cycle number is three. Not every feature in the North Radical has a cycle number of three, however. For example the Head of the North Radical has a cycle number of 6 and the North and South Radicals on Radical 3 have cycle numbers of nine. The iteration track for a point just below the North Atom is shown in Fig. 4. It shows a three armed spiral

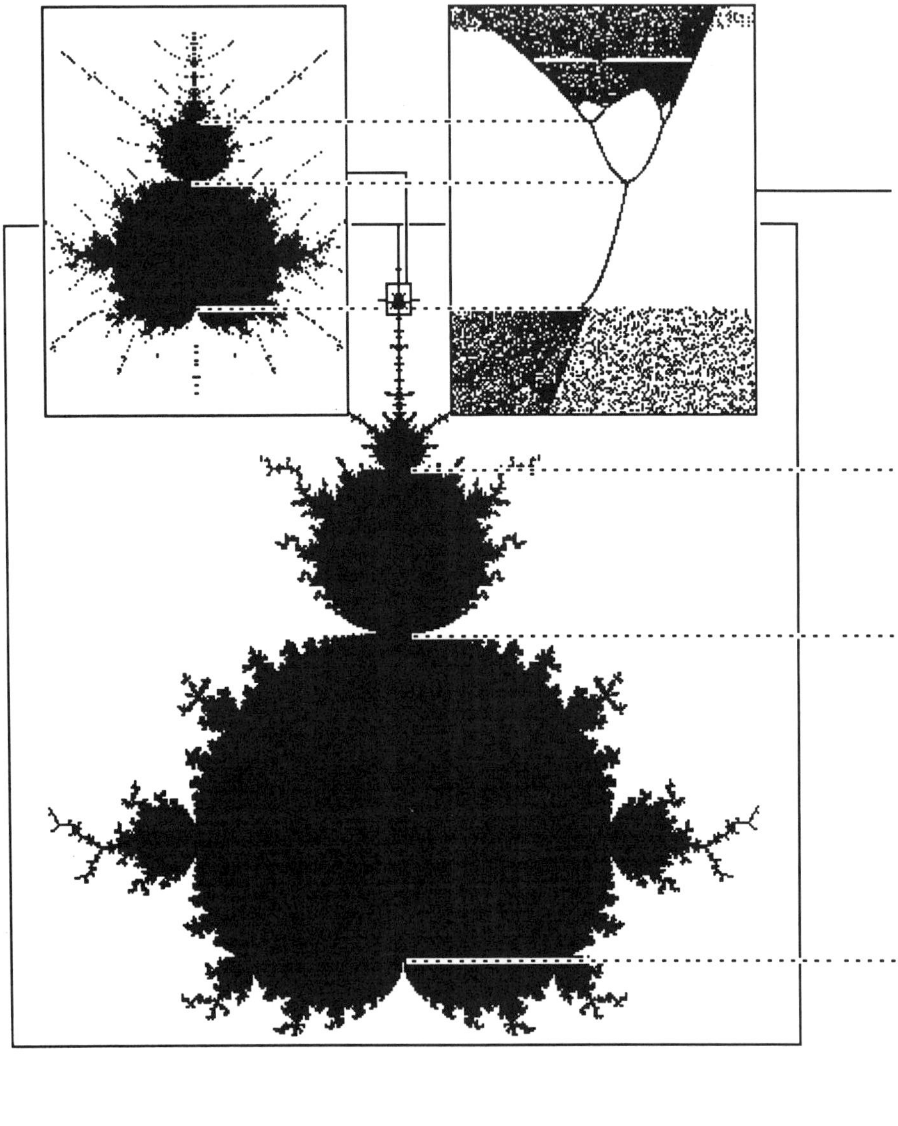

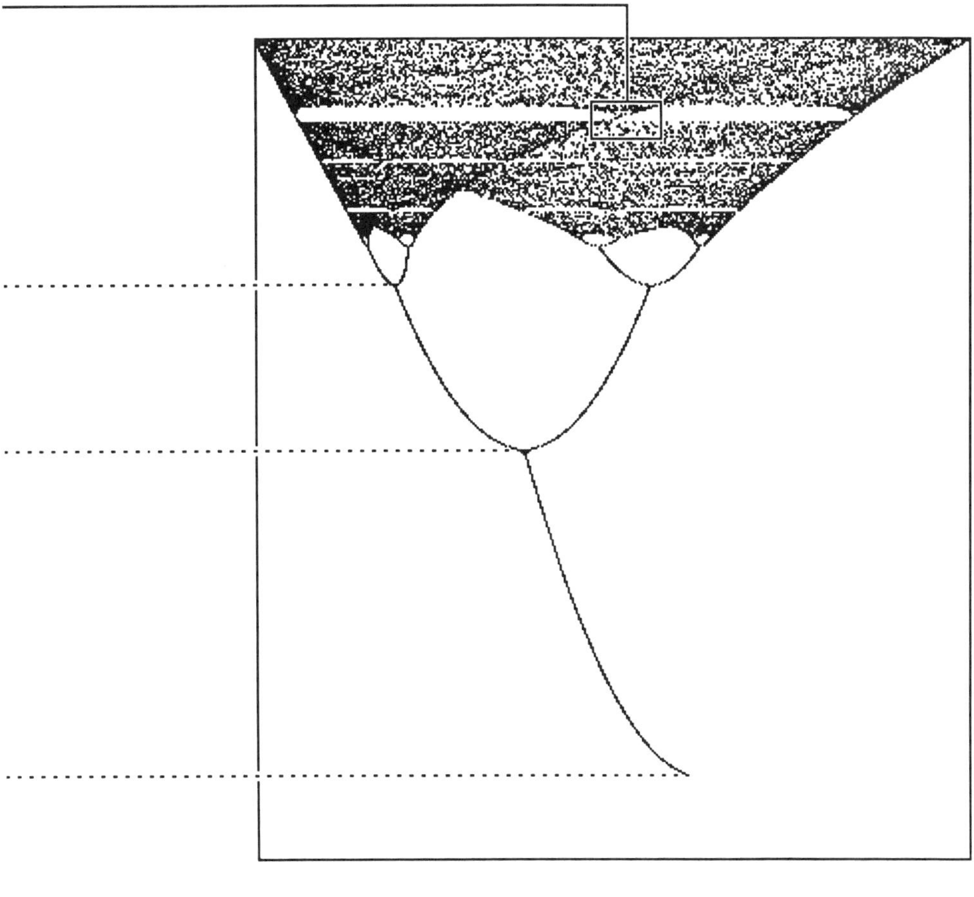

Fig. 2. The Mandelbrot set (upper panel) and the bifurcation diagram (lower panel) with c values on the same scale.

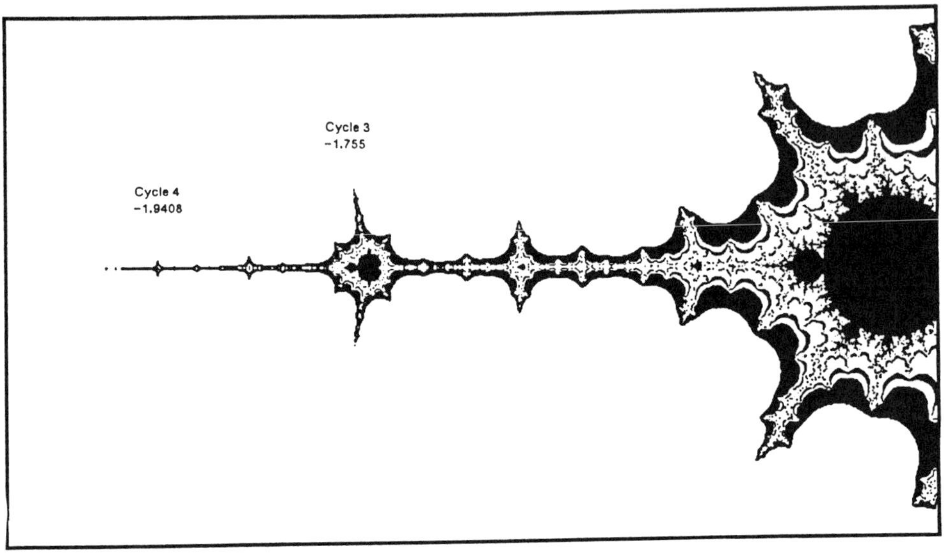

Fig. 3. An enlarged view of the Spike of the Mandelbrot set. The real values are between -2.076 and -1.267. Two of the major midgets (Cycle 3 and Cycle 4) are indicated.

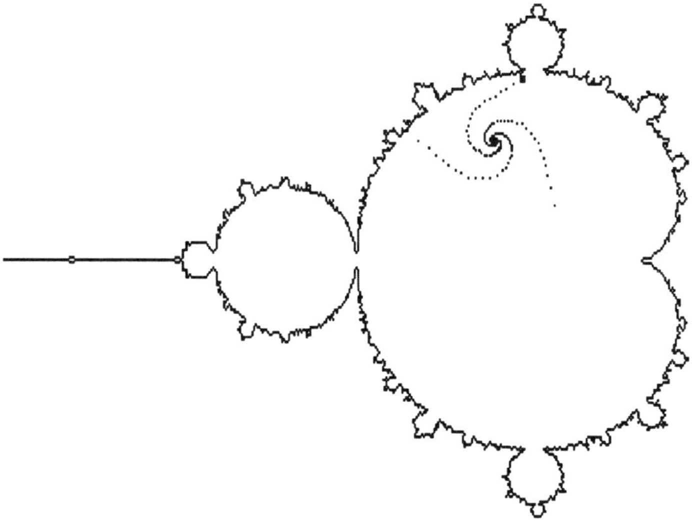

Fig. 4. Iteration tracks for a region just below Radical 3. Since Radical 3 has a cycle number of three, the spiral pattern has three arms.

(three arms because it is formed just below the North Atom which has a cycle number of three) which converges to a single point (since the Body of (**M**) set has a cycle number of one).

If a point c does not belong to (**M**), the iteration sequence will diverge to infinity. To test this divergence, some parameters must be set. The escape radius, er, is a number for which, once the distance from the origin to a point of the iteration sequence exceeds er, we are sure the iteration sequence will diverge to infinity. It is known that er = 2 suffices (e.g. Devaney 1989). The dwell of a point c is the number of iterations necessary for a term of the sequence to first exceed the escape radius. In practice we set a maximum dwell, N, and c is taken to lie in (**M**) if its iteration sequence does not exceed the escape radius in N steps. The escape radius can be set to much larger values than 2, for example if it is set to 100 then the contour lines of constant dwell approximate lines of equal potential, if (**M**) were charged.

In Fig. 5 a three-dimensional representation of the John Dewey Jones midget is

Fig. 5. A three-dimensional representation of the John Dewey Jones midget. The third dimension represents the number of iterations taken at each pixel as it escapes from (**M**).

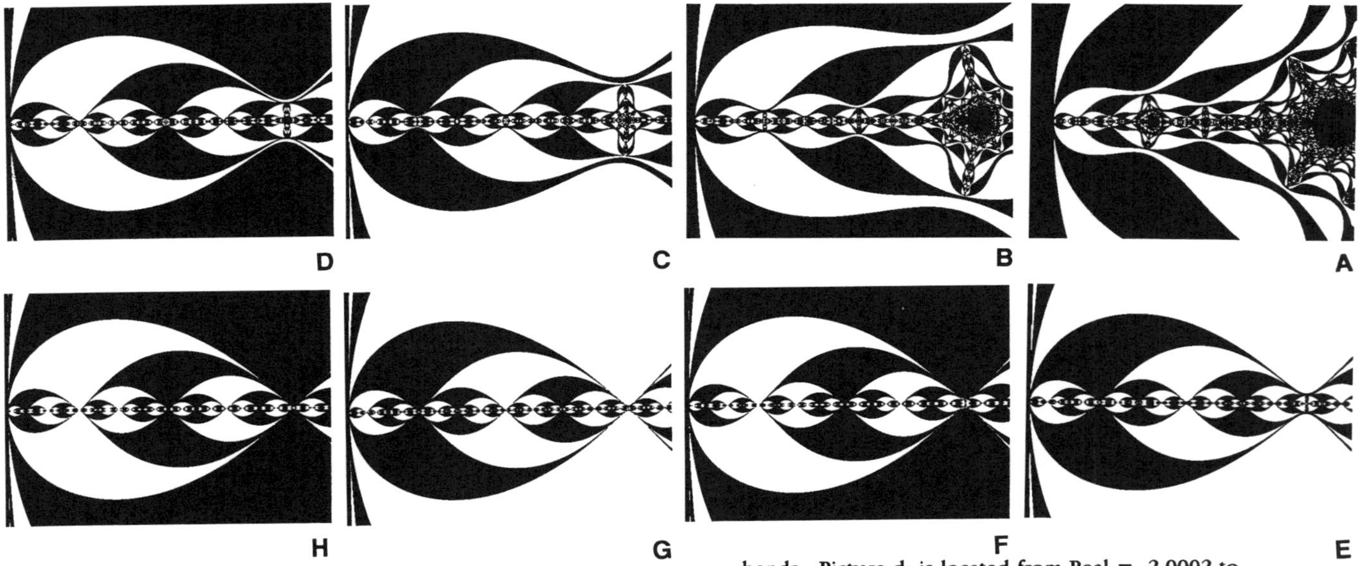

Fig. 6. The Spike of the Mandelbrot Set and enlargements at the end. Picture a: shows the entire Spike (Real = -2.06 to -1.26, Imag. = -0.297 to +0.297, dwell = 150). The prominent midget approximately 2/3 of the way down the Spike is the Cycle 3 midget. Picture b: shows the end of the Spike, from the Cycle 3 midget to the end. Approximately 3/4 of the way from the Cycle 3 midget to the end there is another midget (marked by an arrow) at a pinch in the dwell band - this is the Cycle 4 midget. Picture b: is located from Real = -2.007 to -1.730, Imag. = -0.1037 to +0.1037. Picture c: shows the end of the Spike from the Cycle 4 midget to the end. (Real = -2.0005 to -1.9317, Imag. = -0.0258 to +0.0258). The Cycle 5 midget is at the last pinch in the dwell bands. Picture d: is located from Real = -2.0002 to -1.9831 and Imag. = -0.0064 to +0.0064. The Cycle 6 midget is located in the last pinch of the dwell band. Picture e: is located at Real = -2.00005 to -1.99578 and Imag. = -0.00159 to +0.00159. The Cycle 7 midget is located in the last pinch. Picture f: is located at Real = -2.000012 to -1.998952 and Imag. = -0.000393 to +0.000393. The Cycle 8 midget is located in the last pinch. Picture g: is located at Real = -2.0000026 to +0.000393. The Cycle 8 midget is located in the last pinch. Picture g: is located at Real = -2.0000026 to -1.9997396 and Imag. = -0.0000953 to +0.0001001. Picture h: is located at Real = -2.00000056 to -1.99993628 and Imag. = -0.00002398 to +0.000024196.

shown. This midget is the Cycle 6 midget at R = -1.996376 and was described in Rollo Silver's article in Amygdala, "A Journey to the West" (Namreh 1987). The real axis is horizontal, the axis representing the imaginary numbers appears as if it were coming out of the page and the vertical axis represents the dwell of each point. The black midget is shown as a "mesa" at the top of the figure. In this diagram the maximum number of iterations made was 80, so the mesa represents all points in the figure with dwells = 80 or greater. Other points in the figure exceed the escape radius at various values of the dwell which are smaller than 80. At the bottom of the mesa is a flat, slightly raised yellow area, representing points with low values of the dwell. Close to the midget, the dwell values increase rapidly and it is this steep increase in dwell that creates the effect of the mesa, rising out of the plane. The most interesting areas to investigate in (**M**) are those found close to the black, and these areas are the ones with the highest dwells. We have calculated pictures of areas where the average dwells are 30,000 or more and the iterations in these areas take a long time to complete. However, along the Spike the dwells are quite low. Even at a cycle number of 300 the dwell close to the Cycle 300 midget was only about 3000.

In Fig. 6 a succession of enlargements of the end of the Spike is shown. Fig 6a (upper right) shows a picture similar to Fig. 2. Fig 6b is an enlargement of Fig 6a, from the Cycle 3 midget to the end of the Spike. Each picture in the series is an enlargement (at the same scale as the change from 6a to 6b) of the end of the Spike. A series of envelopes can be seen, enclosing other series of envelopes. These envelopes mark regions of constant dwell. A midget will be found at the common minima of each set of envelope bands if that section of the picture is enlarged sufficiently. The bands, called dwell bands, can be numbered according to the (constant) value of the dwell for points in the band. In Fig. 6a the black area running nearly vertically, at the left, is a portion of dwell band one. Colors can be assigned to each dwell number and one will obtain a series of contours, one inside the other. The pixels from which the iteration sequence exceeds the escape radius in one step are in dwell band one in Fig. 6. The adjacent white area is a portion of dwell band two. The next black area is a portion of dwell band three and the depression in it points to the Cycle 3 midget. In Fig. 6b the areas can be identified in a similar manner and this time the white area (dwell band four) points to the Cycle 4 midget. In a similar manner, in Figs. 4c to 4h the depression in the dwell band (at the end of the Spike) points to the Cycle 5, 6, 7, 8, 9 and 10 midgets. This monograph is concerned with the series of midgets found at these depressions in the dwell band structure. The smallest midget calculated at this time is the one found at dwell band 300, the Cycle 300 midget. The magnification for this image is 1.45×10^{359}, a number so large that it is difficult to comprehend. The Cycle 28 midget, with a magnification of 3.5×10^{31} has an astronomical analogy. If the size of (**M**) is imagined to be the size of the Milky Way galaxy, then the midget represents the size of a hydrogen atom. When we did the Cycle 37 midget, with a magnification of 2.5×10^{42} the analogy became the following. If you imagine (**M**) to be the size of the observable universe then the midget is the size of a proton. As the magnifications increased, we ran out of astronomical analogies! Pictures of each midget computed in this series will be found in Appendix I and at the bottom of each figure will be found the position of the midget on the Spike, the magnification, the size (in pixels) of each picture, the maximum dwell value used and the computer used in the computation.

A

B

C

D

Fig. 7. Locating the Cycle 4 midget. The start of this series of pictures is Fig. 6b. In that picture the Cycle 4 midget is located in the last compression of the dwell bands (Dwell band 1 is the nearly vertical white band at the far left, then band 2 is black, band 3 is white and band 4, with the compression pointing to the Cycle 4 midget, is black). In Fig. 7b the cross is magnified, in 7c the center of 7b is enlarged so that now the eight-armed cross is clearly seen. In 7c the Cycle 4 midget is clearly seen inside a sixteen-armed cross. This picture is similar to the picture on page 48.

The locations of the pictures are as follows:

7a - Real = -1.99677 to -1.99594 and Imag. = -0.00033 to +0.000282.

7b - Real = -1.99654 to -1.99621 and Imag. = -0.00012 to +0.00012.

7c - Real = -1.99641 to = -1.99634 and Imag. = -0.000026 to +0.000023.

7c - Real = -1.99641 to = -1.99634 and Imag. = -0.000026 to +0.000023.

7d - Real = -1.996392 to -1.996346 and Imag. = -0.000012 to +0.000010.

In each picture the dwell was set at 150.

2.1 Last Midgets

If (M) is plotted using an escape radius equal to two and the region near the end of the Spike at -2, +0i is examined in detail (See Fig 6 a-h.) one sees a pattern of lobes of ever-decreasing size. At the left edge of each picture there is a final lobe which includes the point -2, +0i, and at the compression in the lobe which separates that lobe from the preceding lobe in the same dwell band there is a midget. We are concerned with these "last midgets" as a function of dwell band number - and note as well that the cycle number of these midgets is equal to the dwell value of the dwell band whose last compression locates the midget. We have taken this as an empirical rule, expressing the relationship between the dwell band and the cycle number.

Empirically it was found that successive last midgets with cycle numbers increasing by one have sizes smaller by a factor of 16, and their distance from the point -2, 0i decreases by a factor of 4. (See Chapter 9.1) The precise value of these ratios are slightly different from 16 and 4, and, as will be shown in Chapter 3, the distance ratio is asymptotic to 4.0 and the size ratio is asymptotic to 16.0 as the cycle number increases.

In practice, before the cycle-finder program (see Chapter 9.2.3) was developed, a pragmatic method was used to locate the next midget in the series. To compute the position of the next midget from the location of the previous midget one finds a point 1/4 as far from -2, 0i as that of the midget just plotted. The new magnification will be 16 times the old magnification, but first one decreases the magnification slightly and the new location is plotted. The next midget will appear on the screen, and the real coordinate can then be fine-tuned to center the midget and the magnification can be set 16 times the magnification used for the previous midget. From investigations with Freeman's program, V63MBROT (see Chapter 9), it has been noticed that at any selected magnification, the westernmost midget that can be seen is always one of the cycle-n series, so this becomes a pragmatic way of locating such midgets. Magnify the end of the Spike to whatever level you wish and then search for the westernmost midget that is resolvable at that magnification. It will be one of the series.

2.2 Midget Details

Figs. 7a-d show a zoom into the location of the Cycle 4 midget. In Fig. 7a the dip in dwell band four is shown and one can see a four-armed cross at the center. In Fig. 7b the cross is magnified and an eight-armed cross can be seen at its center. In Fig. 7c the eight-armed cross is magnified and now a small midget can be seen at its center. In Fig. 7d the midget is magnified and one can discern the 16-armed cross and inside that the 32- and 64-armed crosses can be seen.

Figs. 8a-h show details in the upper arm of the cross of the Cycle 4 midget. Figs. 8a-e are successive blowups of the center of the previous picture. In Fig. 8e, the center of the picture is taken from the lower feature in Fig. 8e. Figs. 8g and 8h are enlargements of the center of Fig. 8f. Similar sets of lobes can be seen, but Fig. 8d shows a different sort of center. On either side of the main cross at the center can be found replicas of the elongated central feature. In the last panel (Fig. 8h) a new midget can be seen.

Figs. 9a and 9b show the Seahorse Valley of (**M**). Fig. 9b shows an enlargement of the simple contour lines in the upper part of the valley. These pictures can be contrasted with Figs. 10a and 10b which show the Seahorse Valley of the Cycle 7 midget. The area shown in black is the same but the contour lines now have become very complex. Fig. 11 shows a Seahorse Valley from the Cycle 70 midget and now the ray structure in the contour lines has become even more complex. One can see the radicals on the right side of the valley to the lower right. There are three places where the number of spokes doubles as the contours approach the side of the valley.

Figs. 12 - 14 show a series of pictures centered on the Cycle 13 midget, at Real = -1.999999779397. In Fig. 12 one can see 8 rays on the outside and then a small inner circle where the number of rays doubles to 16. In Fig. 13 this circle of 16 is enlarged so that the next doubling can be seen, the circle of 32 rays. In Fig 14 the circle of 64 rays can be clearly seen and the circle of 128 rays can be discerned but the rays are not clear.

2.3 What is Reality in the Mandelbrot Set?

When a program is set to generate a picture of some portion of (**M**) it is necessary to set certain parameters. The dwell and the escape radius have been defined earlier. Also, we mentioned that if a point of an iteration sequence is farther than two from the origin, the sequence will diverge to infinity. The midget pictures in Appendix I were each calculated with an escape radius = two. As mentioned earlier, programs assign colors to each pixel depending on which iteration of the z value exceeds the escape radius. The dwell band interval is the number of iterations that are colored the same color. It can be set to one, or 10, or 20 or any value that one wants. All of these choices affect the way the picture looks, so then the question arises; what is reality in the Mandelbrot set? Fig. 15 shows four versions of the Cycle 85 midget. The dwell in each picture is 2000 and the resolution is 640x480 pixels. What has been changed is the dwell band interval. In Fig 15a it is 1, in 15b it is 11, in 15c it is 21 and in 15d it is 27. The patterns in the pictures vary considerably. In each picture there is an oval, near the upper and lower edge, that is in the same place. This is one of the spoke doubling regions. The number of spokes in the inner oval is twice that of the number of spokes in the outer oval. If one counts in 5 ovals in Fig. 15a the next spoke doubling region is encountered and there are similar places in Figs 15b, c and d. Other ovals in these figures are caused by the cycling of colors through light and dark zones, giving rise to "temporary" ovals. The positions of these ovals will change as the dwell band intervals are changed. Other features can be detected which are the result of aliasing. For example, the regions to the right of East Valley and above the North Radical show structures that would not appear in a picture made at a higher magnification. How one makes a picture of a portion of the Mandelbrot set thus affects the way the picture appears. The region in black, if the dwell is sufficiently high, will be the same, but the details surrounding the black will change from picture to picture.

Strictly speaking, real pictures of (**M**) do not involve dwell bands. Using the standard method with dwell bands suppressed (that is, paint a pixel black if it tests in (**M**), and white if not), the major features are visible but many details disappear. The pictures on the inside cover of this volume contain black and white pictures of (**M**) of this type. The

Fig. 8. Details of a tendril above the Cycle 4 midget. The start of this series of pictures is a low magnification picture of the cycle 4 midget. Figs. 8b - 8e show successive magnications of regions in the tendril. Fig. 8f is an enlargement of the lower part of Fig. 8e. Fig. 8g is an enlargement of the center of Fig. 8e. Fig. 8h is an enlargement of Fig. 8g. A small midget can be seen in the center of Fig. 8h.

The locations of the pictures are as follows:

8a - Real = -1.9596 to -1.9212 and Imag. = -0.0143 to +0.00243.

8b - Real = -1.9458 to -1.9362 and Imag. = -0.00087 to +0.00796.

8c - Real = -1.94233 to -1.94042 and Imag. = +0.00278 to +0.00418.

8d - Real = -1.94153 to -1.94116 and Imag. = +0.00334 to +0.00364.

8e - Real = -1.941242 to -1.941255 and Imag. = +0.003413 to +0.003538.

8f - Real = -1.941382 to -1.941332 and Imag. = +0.003434 to +0.003471.

8g - Real = -1.941363 to -1.941313 and Imag. = 0.003458 to +0.003495.

8h - Real = -1.941348 to -1.941329 and Imag. = 0.003470 to +0.003484.

In each picture the dwell was set at 150.

A

B

Fig. 9. 9a (upper) shows the Seahorse Valley of (**M**). 9b (lower) shows details in the contours, in this case smoothly curved lines.

A

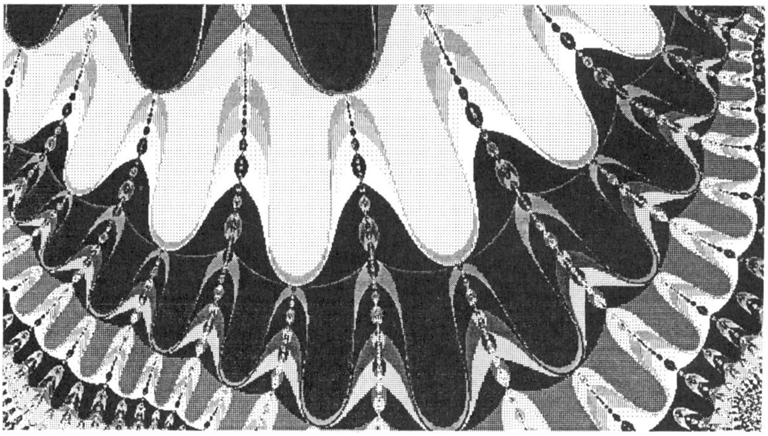

B

Fig. 10. 10a (upper) shows the Seahorse Valley of the Cycle 7 midget. 10b (lower) shows details in the contours, in this case a complicated series of spokes, doubling from one contour interval to the next.

main reason for the disappearing of some features is the multitude of small, wispy hairs around (**M**) (Milnor 1989). A pixel may contain points of (**M**) and yet the test point of the pixel might not lie in the set. The standard method would paint the pixel as not belonging to the set, and so some small features are missed.

The Distance Estimator Method (see page 196 for algorithm and page 195 for figure 4.16 of Peitgen and Saupe 1988) produces pictures which account for these wispy hairs. While this method produces pictures of the "real" set, it does not provide clues for locating some small features. On the other hand, as Fig. 5 indicates, the pattern of dwell bands can help navigate to interesting places much too small to see at the current magnification. Dwell bands point out an interesting distinction between what is "real" and what is "useful."

In Figs. 16a-d we show (**M**) plotted according to four different routines. Fig 16a shows a "normal" mapping with the shades of gray assigned to equal dwell-band intervals. In Fig 16b a logarithmic mapping is used (the color assignment is modified by log(N)/log(max iteration #)). In Fig 16c the distance estimator method is used and one can see the effect of widening in the very narrow tendrils which connect the parts of the M set. Compared to the tendrils coming off the radicals in Fig. 16a the tendrils in 16c are much stronger. In Fig 16d a decomposition method is used. Here the color of each pixel is colored according to which quadrant of the complex plane the final value of z is in when the decision is made if the point is in or out of (**M**). In each of the four pictures the Body, Heads and Radicals of (**M**) are the same but the details surrounding the set vary.

2.4 Midgets are like Snowflakes

Fig. 17 shows what looks like two Mandelbrot sets, one above the other. The top picture is the Mandelbrot set; the lower picture is of the Cycle 3 midget on the Spike. The East Valleys and the Seahorse Valleys have been lined up and it can be seen that the line through Seahorse Valley' of the Mandelbrot set passes slightly to the left of the Seahorse Valley of the Cycle 3 midget. This illustrates one of the distortions present in midgets. The Spike midgets, however, are symmetric vertically. The upper half will reflect across the real axis. Fig. 18 shows two midgets taken from tendrils off Radical 3 and Radical 18. These are the largest, warped midgets in the first tendril. Here it can be noticed immediately that the upper and lower parts of the midget no longer will map into each other. The effect is small for the midget above Radical 3 and much stronger for the midget above Radical 18 (in East Valley). In each case the South Radical is larger than the North Radical.

We have set up two distortion measures. The first measure compares the ratio of the head-to-center distance with the N Radical-S Radical distance of the midget being measured with that ratio for (**M**) itself. As shown in Fig. 17, midgets on the Spike have different ratios depending on their cycle number and position on the Spike. A plot of these ratios for some midgets on the Spike is shown in Fig. 19 (upper panel). The second measure is a measure of angles and is a measure of the distortion seen in Fig. 19 (lower

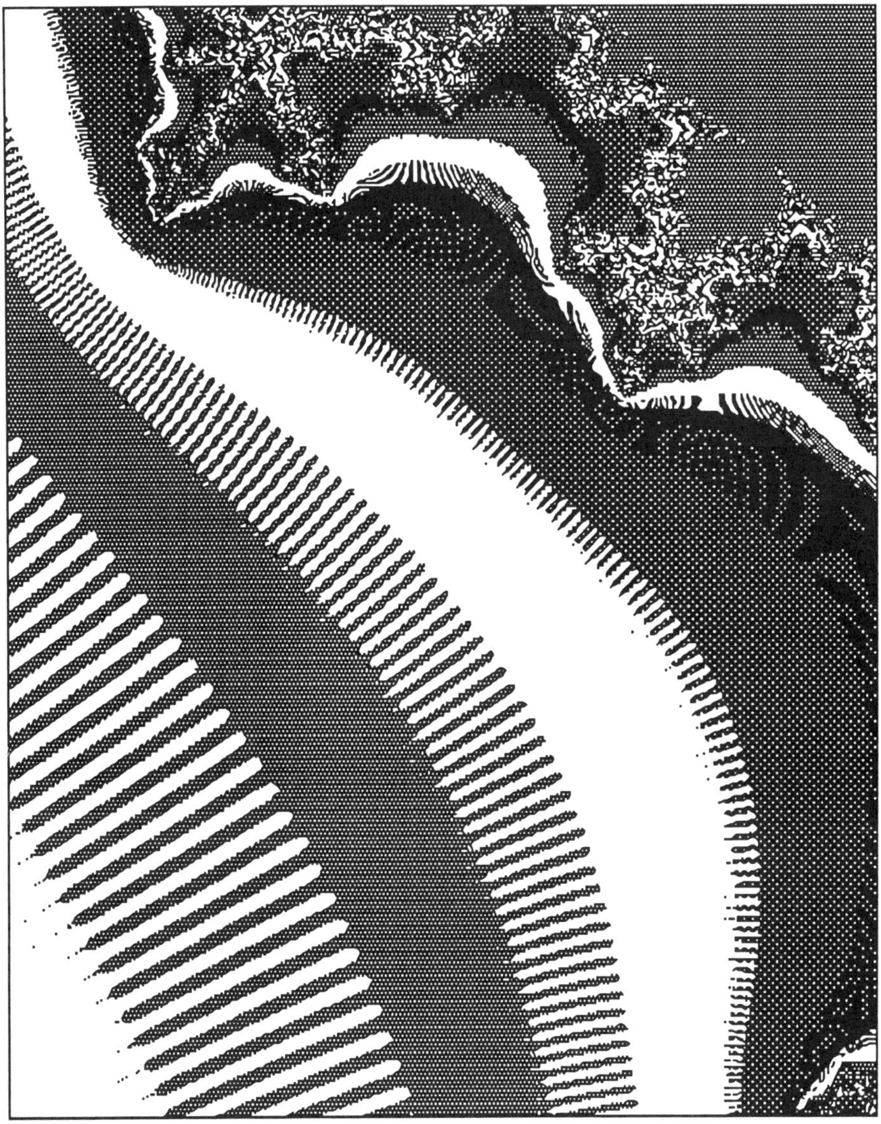

Fig. 11. The Seahorse Valley of the Cycle 70 midget. Now the spokes are extremely crowded.

Fig. 12. The cycle 13 midget at a magnification of $1.8x10^{12}$.

Fig. 13. The Cycle 13 midget at a magnification of $0.9x10^{13}$.

Fig. 14. The Cycle 13 midget at a magnification of $2.6 x 10^{14}$.

panel). A line can be drawn from the center of the Head to the center of the cardioid of the Body. Then lines can be drawn from the North Radical to the center of the cardioid and from the South Radical to the center of the cardioid. For (**M**) itself, the two angles at the center are equal and thus their ratio is 1. For "warped" midgets the angles are different and the ratio, defined as theta = S angle/N angle, becomes greater than 1. A plot of theta for some warped midgets is shown in Fig. 19, lower panel.

We have not yet seen two midgets with the same set of distortion measures. We hypothesize that, if one ignores the reflection of the upper half of (**M**) into the lower half, that no two midgets will have the same distortion measures, if these measures are carried out to a sufficient number of places.

It is interesting to note that there are two different types of midgets. There are symmetrical midgets, found only on the Spike (or on Spikes of other midgets) and here the distortion factor is the ratio of the two distances mentioned above. Then there are the warped midgets which are found off the Spike (or off the Spikes of other midgets) and here the distortion measure is the ratio of the angles formed between the Head and center and the N or S Radical. Work is continuing on determining the values of the two distortion measures for midgets on, and off, the Spike. We have made a list of all the cycle n midgets on the Spike of cycles 3 to 11 and we are determining the distance distortion measure for each one.

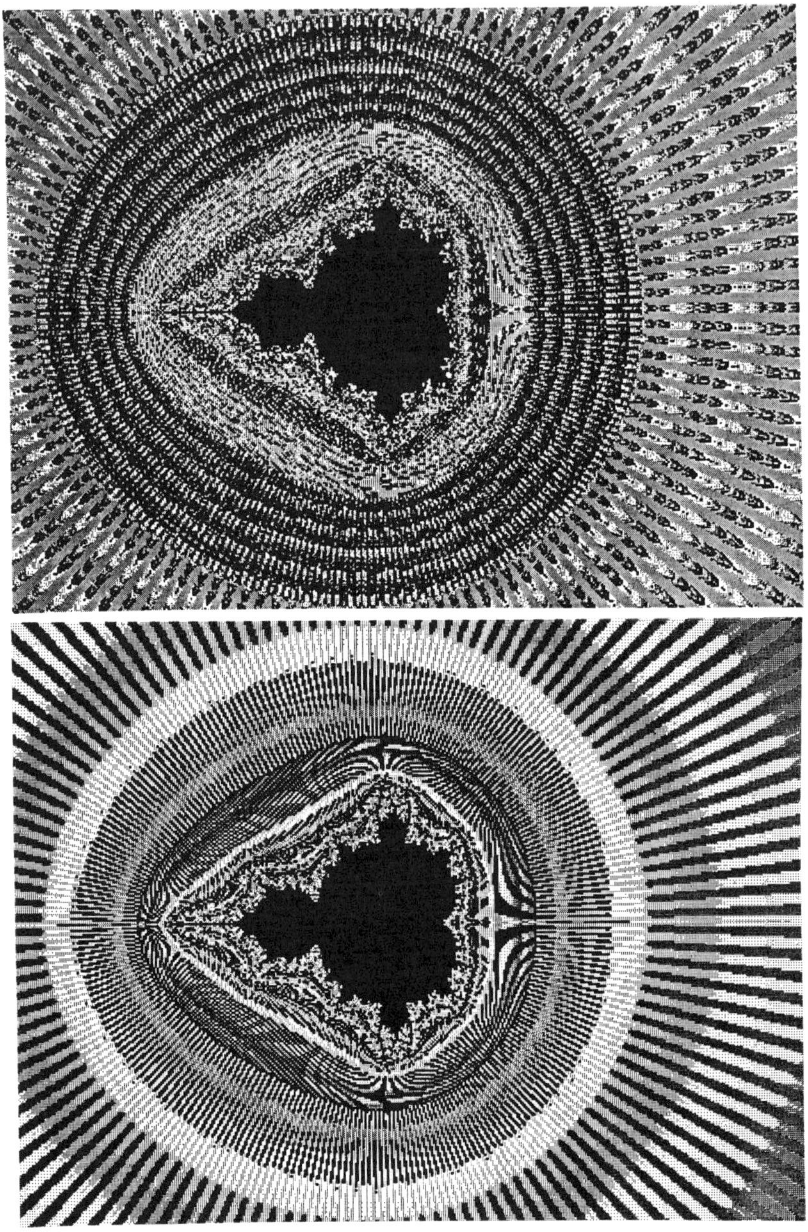

A

B

Fig. 15. Four views of the Cycle 85 midget. The dwell band interval goes from 1 to 11, 21 and 27 in sections a, b, c and d. Fig. 15 a and b are shown above (Dwell bands of 1 and 11).

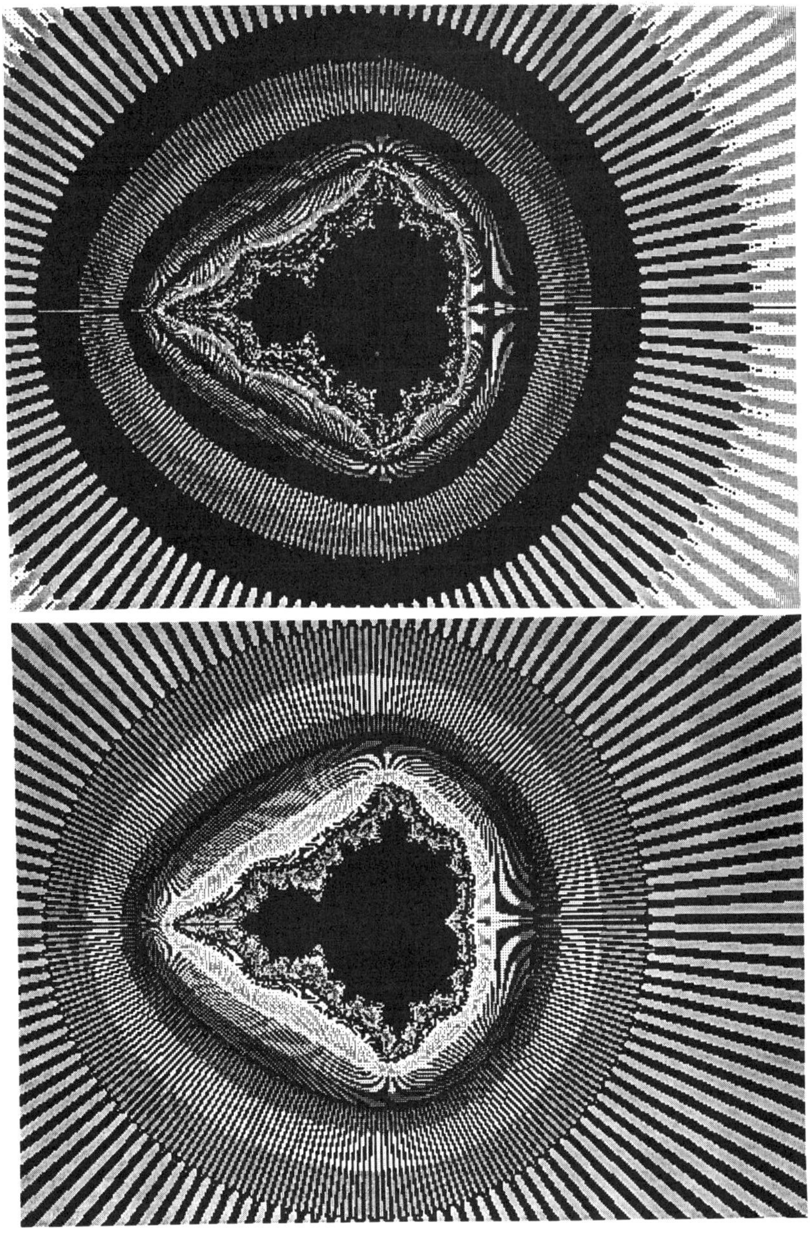

C

D

Fig. 15. Parts c and d are shown above (Dwell bands of 21 and 27).

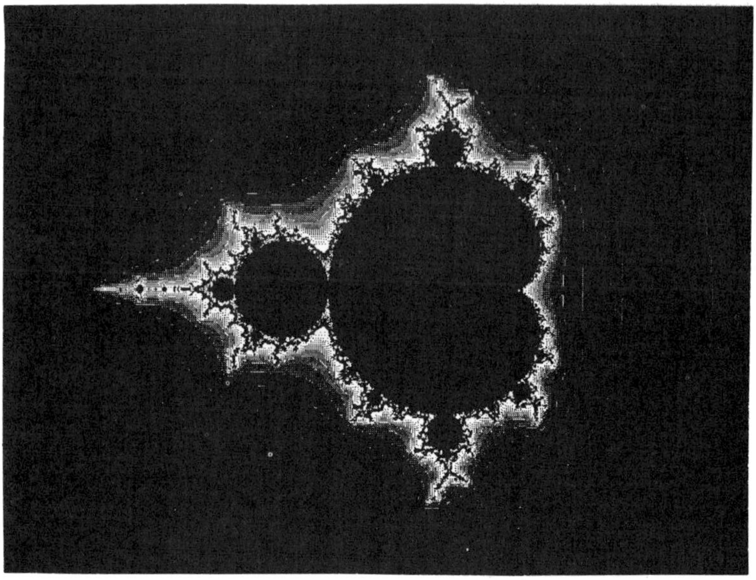

Fig. 16. Four views of (M). 16a shows (M) with "normal" mapping. The dwell in all four pictures was set at 150.

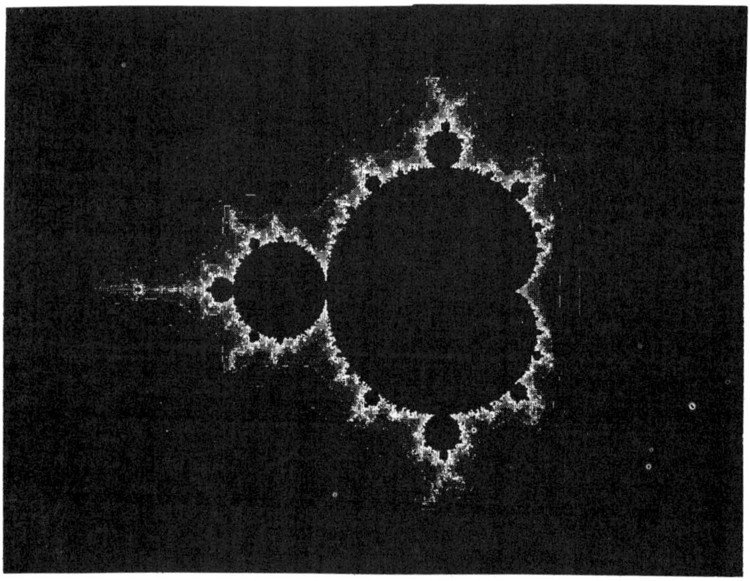

Fig. 16b. (M) plotted with a logarithmic mapping.

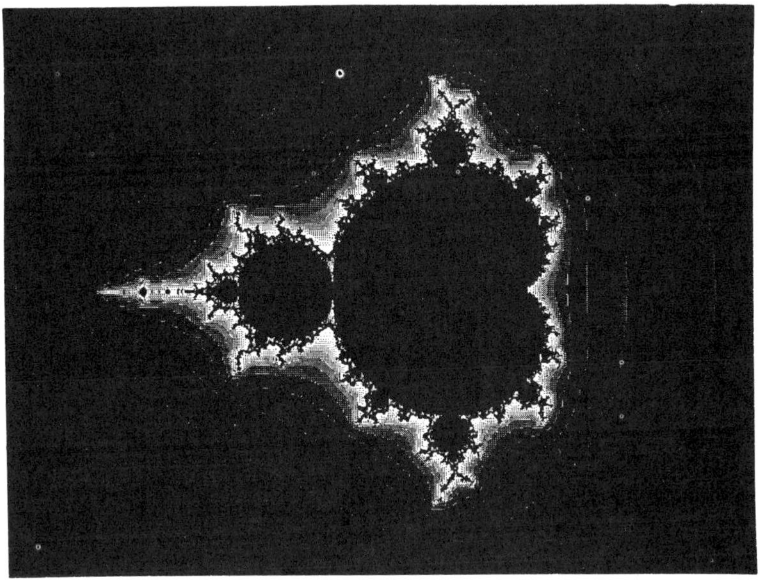

Fig. 16c. (**M**) plotted by the Distance Estimator method.

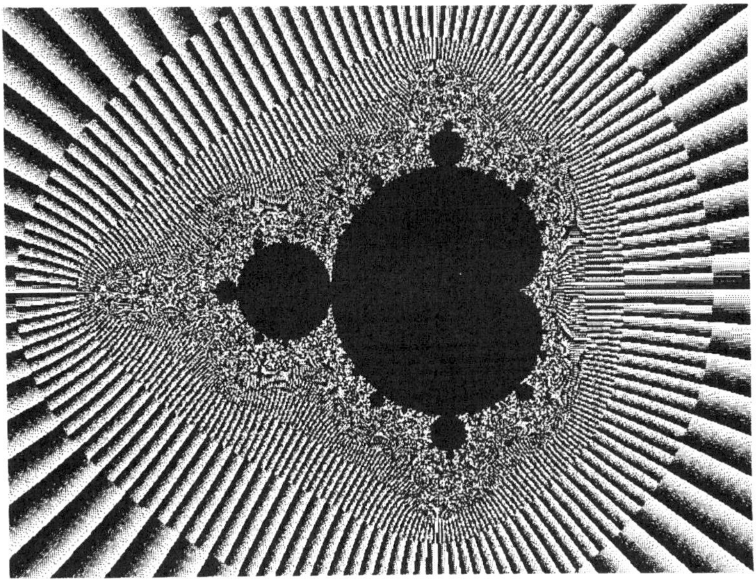

Fig. 16d. (**M**) plotted by the Decomposition method.

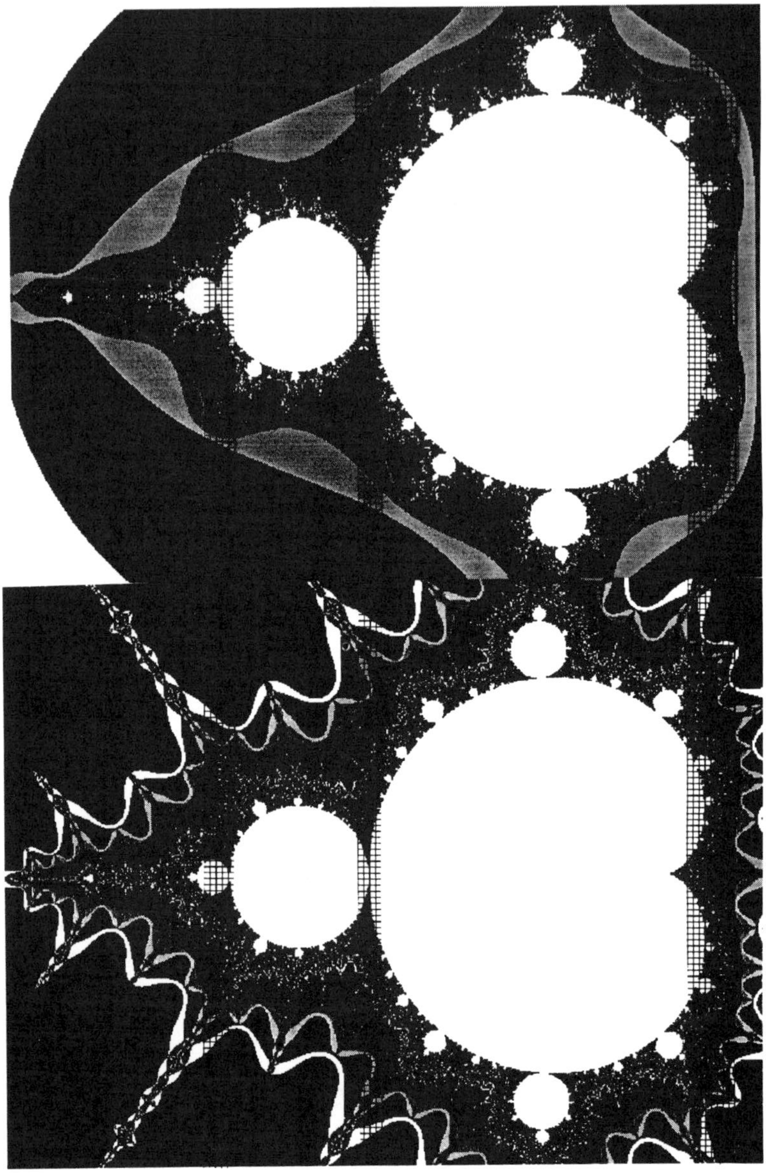

Fig. 17. (**M**) compared with the Cycle 3 midget. The right hand edge of the cardioids are aligned as well as Seahorse Valley. Note, however, that the Head of the Cycle 3 midget falls inside the line drawn from the Head of (**M**). This is an example of the linear distortion found in Spike midgets. Note also that the radicals of the Cycle 3 midget are spaced differently tham in (**M**).

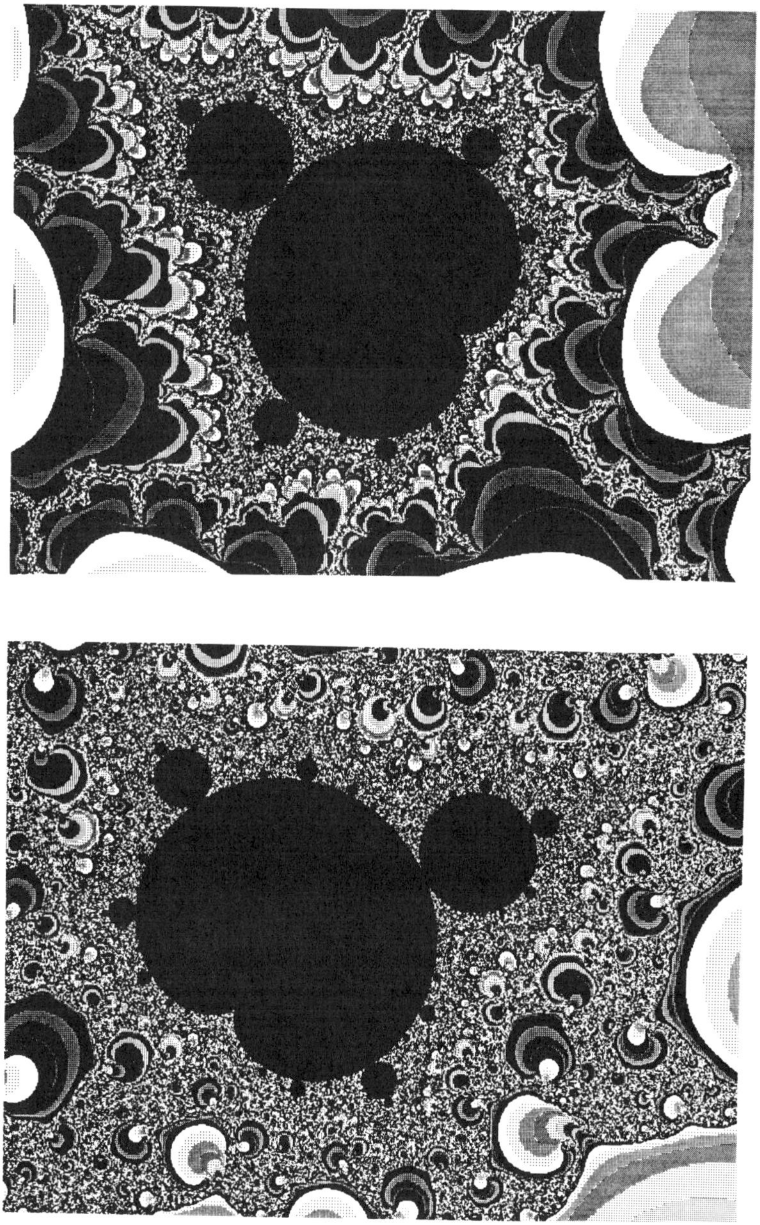

Fig. 18. Two warped midgets. The upper picture shows the large warped midget in the first tendril leading off from Radical 3. The lower picture shows the same feature in Radical 18. In each case the northern part of the midget will not map into the southern part. Note also that the South Radical is larger than the North Radical. This difference is most evident in the case of Radical 18. This is an example of the angular distortion found in off-Spike midgets.

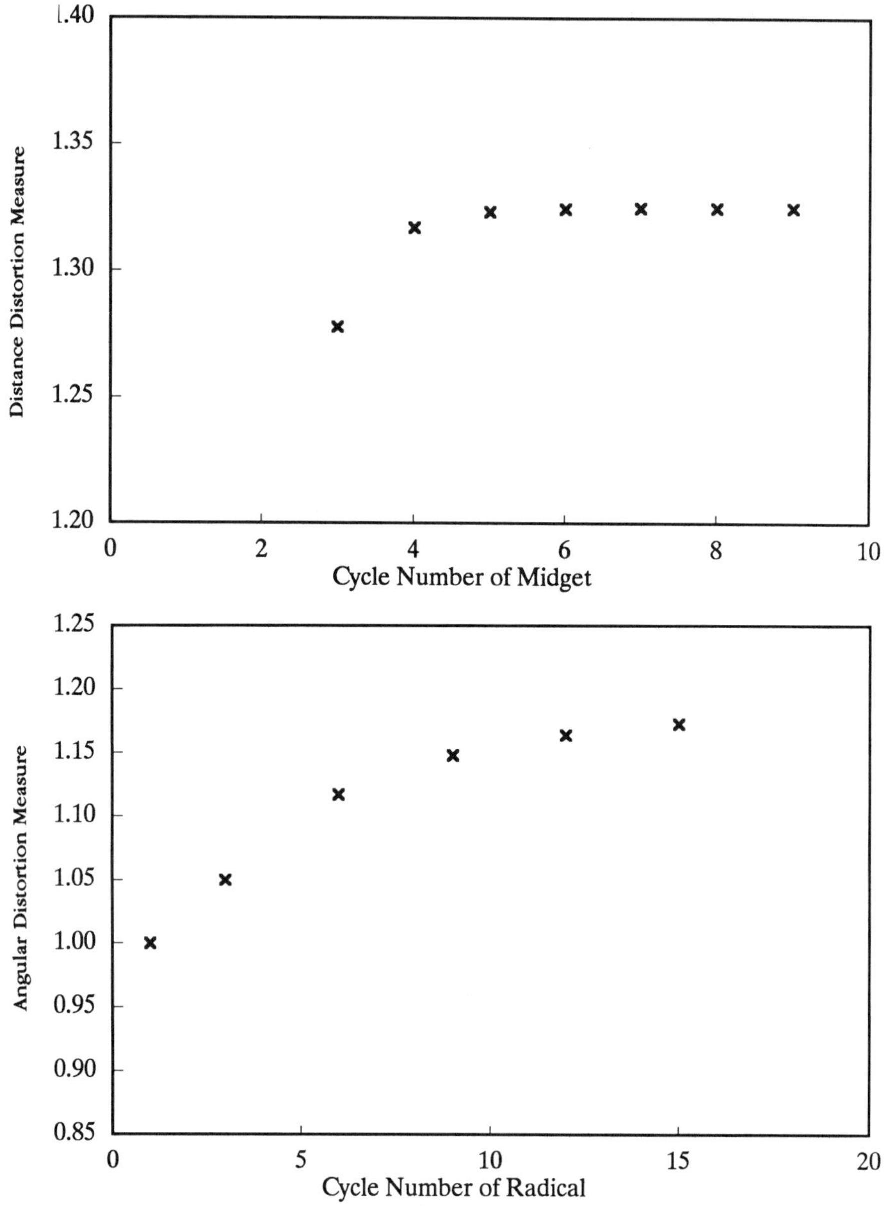

Fig. 19. Distortion Measures. In the upper panel the distance measure distortion (for midgets on the Spike) is plotted versus the cycle number of the midget in the series. In the lower panel the angle distortion measure (for midgets in tendrils, not on spikes) is plotted versus the cycle number of the radical from which the tendril extends.

3. WHY THE MANDELBROT SET IS A NATURAL OBJECT TO STUDY

Pictures of the Mandelbrot set are generated by iterating the function $F_c(z) = z^2+c$. The beauty of these pictures is well-known, but pretty pictures alone are not necessarily interesting. Understanding why the Mandelbrot set is a natural mathematical object to study provides important background for our method of locating midgets, and also is part of the general cultural literacy of those who explore this set.

In this Chapter we present basic notions without the supporting mathematical detail. Some of that detail is outlined in Appendix II.

For a given complex number c, the filled-in Julia set, K_c, is the set of all z_0 for which the sequence

$$z_0,\ z_1 = F_c(z_0),\ z_2 = F_c(F_c(z_0)),\ z_3 = F_c(F_c(F_c(z_0))),\ \ldots \qquad (1)$$

remains bounded for all n. As an example, consider K_c for $c = 0$. A simple calculation (using the polar co-ordinate representation of complex numbers) shows any z_0 with $|z_0| < 1$ gives rise to a sequence of z_n converging to 0. For z_0 with $|z_0| = 1$, we see $|z_n| = 1$ for all n - that is, starting on the unit circle, points stay on the unit circle and so do not diverge to infinity. On the other hand, another polar co-ordinate calculation reveals z_n diverges to infinity if $|z_0| > 1$. We conclude that K_0 is the unit disk $D = \{z_0\colon |z_0| \leq 1\}$.

As another example, consider $c = -2$. By comparing $F_{-2}(z)$ with $F_0(z)$ through a change of variables, we can show K_{-2} is a line segment, the part of the real axis between -2 and 2. (See Appendix II.) These two examples are the only K_c for which we know the precise analytic form; presumably in all other cases the Julia sets are fractals. Can we say anything about these fractals?

Though computing pictures of Julia sets for c selected more-or-less randomly across the complex plane reveals a wonderfully complicated array of shapes, some patterns do emerge. For example, for any c far enough away from the origin, K_c is a disconnected dust of points, a collection of infinitely many infinitely small islands. Magnifying an island always shows the island is itself composed of still smaller islands. Julia sets of this sort are a type of Cantor set, relatives of the Cantor "middle thirds" set familiar to those studying fractals.

Continuing our sampling of points of the complex plane, for some c near the origin, K_c appears to be a mangled circle; for other c, K_c is a collection of infinitely many attached loops, each with infinitely many still smaller loops attached to it, and so on; for still other c, K_c is a tree with infinitely many branches, each of which has infinitely many smaller branches, and so on. What we never see is K_c consisting of two pieces, or three pieces, or four pieces. As far as the number of its pieces, it seems K_c has only two possibilities: it is a Cantor set consisting of infinitely many pieces, or it is a connected set consisting of one (sometimes quite complicated) piece. Are these really the only two possibilities?

In fact, they are. The careful study of these sets was begun early in the 20th century by two French mathematicians, Pierre Fatou and Gaston Julia. Applying powerful tools from complex analysis, they proved theorems having the following applications for the function $F_c(z) = z^2 + c$.

(a) The filled-in Julia set K_c is either a Cantor set or is connected.

(b) Which possibility occurs depends on the behavior of the sequence (1), starting with $z_0 = 0$. If the sequence diverges to infinity, then K_c is a Cantor set. If the sequence does not diverge to infinity, then K_c is connected.

Since K_c is either connected or a Cantor set, a natural mathematical question is to ask for some picture, description, or characterization of the collection of all the c for which K_c is connected. This is the Mandelbrot set, (**M**).

A way not to draw a picture of the Mandelbrot set is to draw K_c for a lot of c and then determine which K_c are connected. Of course, this could work in principle, but implementing such a scheme would be extremely difficult and slow. A method which will work, and is often used, is suggested by (b). Divide a region of the complex plane into little squares (pixels), select a c from each square (the center or the northwest corner are common choices), and generate the sequence (1) starting from $z_0 = 0$. If the sequence does not diverge to infinity, then c belongs to the Mandelbrot set and typically the pixel is painted black. After testing each pixel of the region, the collection of black pixels forms an approximation of the part of the Mandelbrot set in that region. (The wispy, filamentary nature of the boundary of the Mandelbrot set makes this sometimes a crude approximation: filaments of the Mandelbrot set can pass through a pixel without encountering the c selected from that pixel. In fact, John Milnor (Milnor 1991) reported that Shishikura has proved that the boundary of the Mandelbrot set has Hausdorff dimension 2, so on a very small scale the filaments become extremely crowded.) To tell if the sequence (1) is going to diverge to infinity, we rely on a calculation (Prop. 3.2.4, Devaney 1989) showing divergence occurs if some z_n is farther than 2 from the origin. The main problem in producing pictures of (**M**) arises in determining how many elements of the sequence need to be generated to tell if the sequence will eventually diverge. For this problem there is no sure answer; in practice we select a maximum number, N, of iterates (called the maximum dwell), and if the first N terms of (1) have not gone farther than 2 from the origin, we say the terms of (1) will all remain bounded and the corresponding point c belongs to (**M**). In general, the larger the dwell, the better the resulting picture approximates (**M**). Of course, the larger the dwell, the longer it takes to generate the picture. Points outside (**M**) are colored according to the number of terms of the sequence we must generate before finding one farther than two from the origin.

4. A METHOD FOR LOCATING SOME FEATURES OF THE MANDELBROT SET

The main features of (**M**) observed so far are (at least approximately) cardioids and disks. Associated with each of these is a unique point, called the center. (See Appendix II for the definition of the center.) Taking c to be one of these centers, the sequence (1) is repeating - that is, there is some number n for which $z_n = z_0$, and consequently $z_{n+1} = z_1$, $z_{n+2} = z_2$, ... Call this n the cycle number of the disk or cardioid with center c. As we show in Appendix II, the centers of cycle number n are the roots of a polynomial of degree 2^{n-1}. Here are the first few of these polynomials.

$n = 1 \qquad f_1(c) = c$

$n = 2 \qquad f_2(c) = (f_1(c))^2 + c = c^2 + c$

$n = 3 \qquad f_3(c) = (f_2(c))^2 = (c^2 + c)^2 + c$

$n = 4 \qquad f_4(c) = (f_3(c))^2 + c = ((c^2 + c)^2 + c)^2 + c$

...

All the others can be generated inductively by the formula

$$f_{n+1}(c) = (f_n(c))^2 + c \qquad (2)$$

That is, knowing $f_4(c)$ we can find $f_5(c)$, knowing $f_5(c)$ we can find $f_6(c)$, and so on.

Consider some examples. There is only one root of $f_1(c) = 0$, namely $c = 0$, and this is the center of the main cardioid of the Mandelbrot set. Next, observe

$$0 = f_2(c) = c^2 + c = c(c + 1) \qquad (3)$$

So $f_2(c)$ has two roots, $c = 0$ and $c = -1$, but of course $c = 0$ has already occurred as a root of $f_1(c)$. The root $c = -1$ is the center of the head of the Mandelbrot set. Before making conjectures about any patterns, we look at two more examples.

The cycle 3 centers are solutions of the equation

$$0 = f_3(c) = c^4 + 2c^3 + c^2 + c = c(c^3 + 2c^2 + c + 1) \qquad (4)$$

That is, $c = 0$,
$c = -1.7548777$, and
$c = -0.12256177 \pm 0.74486177i$.

Of course, $c = 0$ is the center of the main cardioid. The other real root is the center of the main cardioid of the largest midget (the Cycle 3 midget) along the Spike, and the complex roots are the centers of the north and south atoms (cycle 3 disks).

The cycle 4 centers are solutions of the equation

$$0 = f_4(c) = c^8 + 4c^7 + 6c^6 + 6c^5 + 5c^4 + 2c^3 + c^2 + c$$

$$= (c^2 + c)(c^6 + 3c^5 + 3c^4 + 2c^2 + 1) \qquad (5)$$

That is,

$c = 0$,
$c = -1$,
$c = -1.9407998$,
$c = -1.3107026$,
$c = -0.15652017 \pm 1.0322471i$, and
$c = +0.28227139 \pm 0.53006062i$.

The centers $c = 0$ and $c = -1$ we have encountered before (cycle 1 and cycle 2). The root $c = -1.9407998$ is the cardioid center of the Cycle 4 midget along the Spike, $c = -1.3107026$ is the cardioid center of the Head of the Head (a cycle 4 disk), $c = -0.15652017 \pm 1.0322471i$ are cardioid centers of Cycle 4 midgets in the antenna structure of the North and South Atoms, and $c = 0.28227139 \pm 0.53006062i$ are the centers of the Cycle 4 Atoms, the largest disks attached to the main cardioid to the right of the North and South Radicals.

By now a general feature may seem plausible. If c_0 is a cycle m center and m is a factor of n, then c_0 is a cycle n center. That is, the polynomial $f_m(c)$ is a factor of the polynomial $f_n(c)$. (The reason is given in Appendix II.)

To understand this observation, consider the examples we have produced. First, 1 is a factor of 2, 3, 4, ..., and from equations (3), (4) and (5) we see $f_1(c) = c$ is a factor of $f_2(c)$, $f_3(c)$ and $f_4(c)$ - indeed, from equation (2) we see if c is a factor of $f_n(c)$ then c is a factor of $f_{n+1}(c)$. Next, 2 is not a factor of 3, and from equations (3) and (4) we can verify $f_2(c)$ is not a factor of $f_3(c)$. Finally, 2 is a factor of 4, and we see from equations (3) and (5) that $f_2(c) = c_2 + c$ is a factor of $f_4(c)$. These observations motivate us to define a proper cycle n center as a cycle n center which is not a cycle m center for any $m < n$. To find the proper cycle-n centers for small n, we follow three steps.

(i) Generate the polynomial $f_n(c)$ using (2).

(ii) Reduce this polynomial by dividing out the factor $f_m(c)$ for m dividing n.

(iii) Find the roots of the reduced polynomial by some appropriate numerical scheme.

Of course, this is not quite the method we used for finding the high-order cycle centers. A polynomial of degree 2^{100} is not especially manageable; our short cuts (applicable to the particular sequence of midgets we have investigated) will be described in the section discussing the programs.

To locate small features in (**M**), simply picking a region and magnifying repeatedly will yield something, but perhaps only after an awfully long time. We used the roots of the polynomials $f_n(c)$ to locate the centers of midget cardioids. In addition to allowing us to find the locations of very small midgets, certain properties of the polynomials allow us to find midgets in a particular sequence, to be described in a moment. Using them in other ways, these polynomials allow us to locate branch tips and the centers of stars and spirals in the boundary of (**M**).

5. SERENDIPITOUS SCALINGS

The original goal of this project was to find a method of locating extremely small midgets. Before starting a magnification which may occupy several days, we wanted to be certain something of reasonable size would show up in the picture. Taking into account some properties of the polynomials described in Chapter 4, the easiest family of midgets to locate was the cycle n midget farthest to the left along the real axis, for $n = 3, 4, 5, \ldots$. An empirical rule for locating members of this family had been deduced by (KWP); using our polynomials we were able to produce compelling numerical evidence for this rule [Frame, Philip and Robucci 1991]. Denoting by c_n the center of the left-most cycle n midget along the real axis, we observed as n gets large

$$(c_n - c_{n+1})/(c_{n+1} - c_{n+2})$$

approaches the limit 4. In this way, knowing c_n and c_{n+1} we can predict where to look for c_{n+2}, namely

$$c_{n+2} = (5c_{n+1} - cn)/4$$

This relation can be used to navigate down the Spike, hopping from midget to midget in increasing cycle numbers.

Of course, locating the centers of the midgets is not enough for producing a reasonable picture; some notion of the size of the midget is needed as well. As our predictor of the size of a midget, we used the head-to-center distance. Recall for a cycle n cardioid, the head is a cycle 2n disk component, and again using properties of the family of polynomials, we were able to locate the centers of the heads of the midgets in our sequence. The head-to-center distance is the distance from the cardioid center to the head center. As n gets large we observed that

$$(c_n - h_n)/(c_{n+1} - h_{n+1})$$

approaches 16. Consequently, in passing from the cycle n midget to the cycle (n+1) midget in our sequence, magnifying the scale by a factor of 16 gave a picture of approximately the same size as the previous midget.

The scalings represented by these two ratios (reported in more detail in [Frame, Philip and Robucci 1991]) appear to represent a new symmetry of the Mandelbrot set; they were discovered while trying to locate and size midgets along the Spike of the Mandelbrot set. What other symmetries might be lying still unnoticed?

6. HISTORY OF MAGNIFICATION OF MIDGETS ON THE SPIKE

In 1987 Namreh (Rollo Silver) published an article in Amygdala (a newsletter devoted to papers on the Mandelbrot and Julia sets, and related topics) entitled "A Journey to the West". He showed pictures of the biggest midget and the John Dewey Jones midget (referred to earlier). The JDJ midget has a magnification of $1.4x10^5$. (See the pictures on pages 7 and 50.) In 1988, while Mandelbrot was at Union College giving the Steinmetz Lecture one of us (AGDP) asked him what was the largest magnification that had been computed on a mainframe computer. He said the largest magnification that he knew of was an area that had been magnified to Avogadro's number ($6x10^{23}$). KWP then wrote a True Basic program (utilizing a True Basic toolkit that converted long numbers to strings) and he was able to calculate the Cycle 28 midget with a magnification of $3.51x10^{31}$. (See the picture on page 72.) This midget was reported on by A. K. Dewdney (1989) in Scientific American in his column on Computer Recreations.

Using a high precision program written by AR, he and AGDP have continued the investigation of midgets to much higher magnifications. We reached a googol in March of 1990 (see the picture on page 129) and then did the Cycle 175 midget, at a magnification of $4.44x10^{208}$ in June. This picture took almost two weeks of cpu time on a VAX 6220 and a Zenith Turbosport and was reported in an article by Frame et al. (1991). Following this, pictures were made of the Cycle 200, 250 and 300 midgets. The Cycle 300 midget was calculated on a Dec Workstation (the time on which was granted by Lance Spallholz at Union College) and took approximately one month of cpu time to calculate the 100x76 pixel image which can be seen in Fig. 141. It has a magnification of $1.45x10^{359}$. An article concerning the Spike midgets has been published in Amygdala (Philip and Robucci 1991).

The cycle-finder program (AR) finds the position of any cycle-n midget in our sequence and this program was run on one of the Union College VAXs until it had calculated the position of every cycle-n midget up to Cycle 300. This table has been placed on a floppy disk. The midget positions, for those pictured in this volume, along with other pertinent data, will be found in Appendix III.

7. GALLERY OF MIDGETS FROM CYCLE 3 TO CYCLE 300

In this section we discuss pictures of the midgets calculated, from Cycle 3 to Cycle 300. (The picture section will be found in Appendix I). The first 12 pictures were calculated with the program FRACTINT at VGA resolution (640x480 pixels). Cycles 13 through 28 were calculated (at resolutions from 400x300 to 200x150) on the Union College VAXs with the program HIMBROT, written by AR. Cycles 29 through Cycle 300 midgets were calculated at a resolution of 100x75 or 100x76, again with the program HIMBROT.

As one scans the pages of pictures of the midgets of higher and higher cycle number there are some interesting changes to be seen in their features. The Cycle 3 midget has four tendrils extending to the edge of the picture, eight tendrils reach iteration bands five iterations from the edge, sixteen tendrils in the next layer and so on, doubling in each successive layer. In the picture for the Cycle 4 midget, the tendrils have become more prominent and now the 8-, 16-, 32- and 64-rayed structures can be seen. By the Cycle 12 picture the 256-rayed layer can be counted clearly. In the picture for the Cycle 13 midget the resolution drops to 400x300 pixels and now the ring of 128 rays is the innermost ring that can be resolved. For the picture of the Cycle 29 midget the resolution drops to 100x75 pixels and only the ring of 64 rays can be counted. The cpu time taken to compute these pictures increases as numbers of greater length are used and as the magnification increases, thus the deeper pictures are computed using a coarser grid to keep the computation times reasonable. To compute one of the 100x76 pixel pictures at full VGA resolution would take over 40 times the amount of cpu time and for the Cycle 300 midget this would come out to over three years of cpu time on a RISC workstation.

In these low-resolution pictures the midget sometimes assumes an odd shape due to problems of aliasing, mentioned in Chapter 2.3. Imagine placing a coarse grid on top of a complicated, detailed picture. Ascertain the color of the picture at every intersection point and color the entire square above and to the right of that intersection point the same color. One now has a very coarse image which caries information from widely spaced points in the underlying image. So a small change in the position of the grid will produce changes in the appearance of the resulting new picture. If one scans the shapes of the midgets shown in the 100x75 or 100x76 pictures some distinctly odd shapes can be seen as a result of aliasing. For example, the Cycle 52 midget has a very odd, flat-headed appearance and the head of the Cycle 125 midget is quite "boxy". The North and South radicals, small features at this scale, show great changes in their appearance. Sometimes they merge into the body and can hardly be distinguished as radicals. Sometimes the tendril structure appears to the left of the radical, other times the main tendril appears to the right of the radical. The largest midget on the Spike also undergoes great variations on this scale. The Cycle 47 and 67 midgets hardly show any evidence of this feature and the Cycle 50 and 59 midgets show a large 9 or 15 pixel box at the position of the largest midget on the Spike.

If the most highly magnified midgets are compared to those at low magnification a major difference can be seen in the areas surrounding the midgets. In the low-magnification pictures one can see the individual rays, doubling in number at each transition zone. The Cycle 12 midget is a good example. On the outside edges one can see 16 rays. The next layer in is the 32-rayed structure followed by the 64-, 128- and 256-rayed structures. At lower resolutions these transition zones would appear as concentric circles, or ovals, but the individual rays would not be resolved. If one looks at the Cycle 43 midget an oval can be seen, almost tangent to the top and bottom of the picture. This is the transition zone between the 64- and 128-rayed structures. Further in a darker oval can be seen. This is the transition zone between the 128- and 256-rayed structures. Thus these ovals are "real"; they will appear in every picture of the Cycle 43 midget, no matter what the dwell-band settings are. There are other ovals that appear in this picture but they are the result of the cycling of the color map and will change in position as the dwell-band interval is changed. If one remembers Fig. 4 (the three-dimensional plot of the JDJ midget) one can see what is happening. The "mesa", supporting the midget, is oval in shape at the top and more circular at the bottom. As the colors cycle in a black and white picture, there will be zones where a light color is next to a dark color and this will give rise to a pseudo-oval in the picture. The positions of these pseudo-ovals will change from midget to midget, or in a given midget if the dwell bands are changed. However, the major transitions zones (64-128, 128-256) will be stable against changes in color assignment or dwell-band interval.

8. RELATIONSHIPS

We plot, in Fig. 20, the log of the distance between successive centers of cycle-n midgets and log of the head-to-center distance of successive cycle-n midgets from Cycle 3 to Cycle 101 as a function of the cycle number n. The relation is linear over the entire range. This is the range for which we calculated almost every midget in the series. The separated points beyond 100 also fall on the line determined in Fig. 20. Additional data concerning the midgets in the series is presented in Tables 1 and 2. Table 1 shows the cycle number versus the ratio of the distances between the centers of successive midgets. As the cycle number increases the ratio approaches 4 at the limit. In the table entries one can note the increasing number of zeros in each value of the ratio as n increases. Table 2 shows the ratio of successive head-to-center distances for the midgets in the sequence. This ratio approaches 16 as n increases. Again, one can note the increasing numbers of zeros as n increases. There are a few places where the pattern is broken (n = 34/35, 57/58, 86/87 and 95/96). We do not know the reason for these breaks.

Since the midgets were calculated so that each was the same size in the pictures it is possible to investigate the average dwell for each midget by making an average of the dwell values in the first row of each picture. These values have been plotted in Fig. 21 which shows the average dwell as a function of the cycle number, n. The relation is nearly linear, the dwell increasing slightly more than linearly with n. However, the relation can be used to predict the dwell value that must be used in calculating midgets beyond Cycle 300.

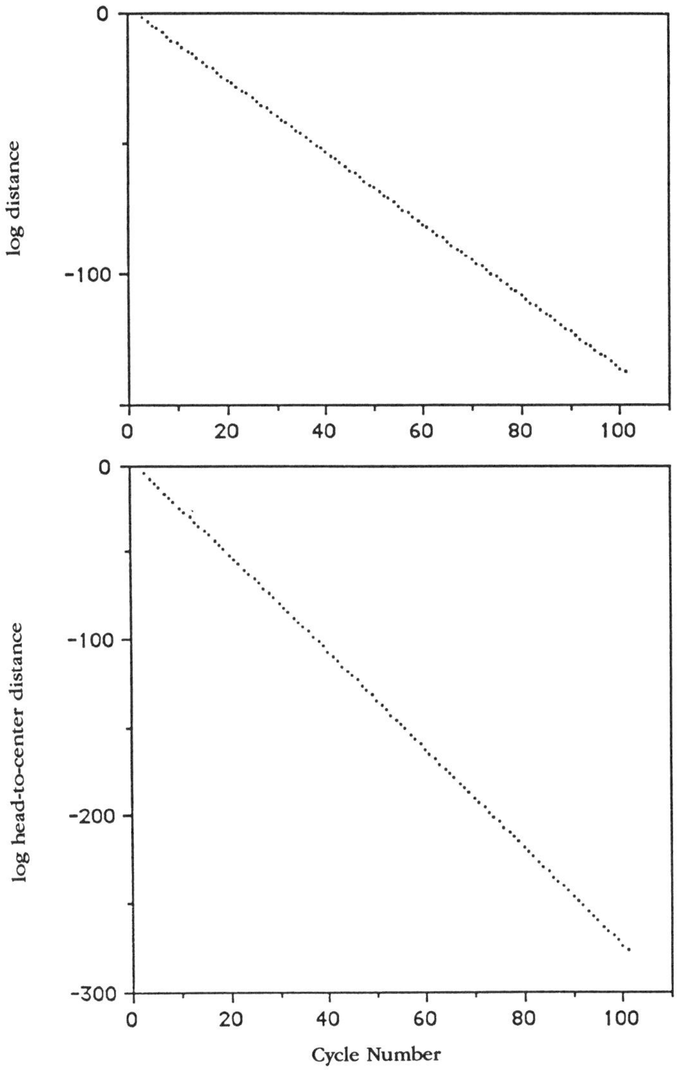

Fig. 20. In the top panel the log of the distance between the centers of successive cycle-n midgets in our series is plotted as a function of the cycle number n, $3 \leq 101$. In the lower panel the log of the head-to-center distance of successive cycle-n midgets in our series is plotted as a function of the cycle number n, $3 \leq n \leq 101$.

Table 1

Cycle number versus the ratio of distances between centers of successive midgets in our series.

n	$\frac{c_n - c_{n+1}}{c_{n+1} - c_{n+2}}$
3	4.16637414605908662208496455654839111377920160972260972260354
4	4.07459062274231735159872047478088055695713550822857469822557
5	4.02710240282901100185576866924130444467989408991256485727581
6	4.008961453998398074383517895835107577789206905695806750914514
7	4.00279737020555418877607870167518113275671966025453046005272
8	4.00083979224262016238306903536659879312871447469506726371456
9	4.00024517934233964400834854096899560219758409417089377370461
10	4.00007011346947268604441935445897870877419505642951991048702
11	4.00001973395248636269474770773131172432979721769628461560051
12	4.00000548496041719596635305681848766516348688043312852899005
13	4.00000150911421351233416983822142941109388727464014583290479
14	4.00000041174754750354060161134376777114411930757412052662773
15	4.00000011155417077780054836089339684144202559220741679612024
16	4.00000003004286630462413131008633668373963446705923881158999
17	4.00000000804929767225869647538780671141504228956751437866251
18	4.00000000214696970508878925854772371226865818738297428307093
19	4.00000000057040375014435997223335468317845104367457196263031
20	4.00000000015101626851953529433021629056983224685242576916949
21	4.00000000003985789976446295084695921763677430746956084488204
22	4.00000000001049043311938054289697531184468438351487206318541
23	4.00000000000275409782443605834183826009929569225117701075806
24	4.00000000000072139684225845777744272378692858709212242616572
25	4.00000000000018856730710209327881315809436431153971247802389
26	4.00000000000004919635090990112683171621201009654984064735376
27	4.00000000000001281271876107028681967591528956519905441238843
28	4.00000000000000333158744866636064592263708822339640222221017
29	4.00000000000000086499880176628997376054102053109799957119195
30	4.00000000000000022427518534149746454013573957615773096013330
31	4.00000000000000005807516756035581694248314990257490858270184
32	4.00000000000000001502038469633424713050425587400643598046215
33	4.00000000000000000388049437565084338365270649189938510756026
34	4.00000000000000000100147314427180282663159623930366237300917
35	4.00000000000000000025820567369677466139102603627653680330808
36	4.00000000000000000006651076531763883981033834237978619610126
37	4.00000000000000000001711752805425865916238117310636739666795
38	4.00000000000000000000440184119477690209311325917845420182901
39	4.00000000000000000000113107509399728484891236447207677886078
40	4.00000000000000000000029042247232508604363690495402606935065
41	4.00000000000000000000007451904278771271876144959821830965622
42	4.00000000000000000000001910811687353848165341955148168635444
43	4.00000000000000000000000489661826253719590411926222864384583
44	4.00000000000000000000000125405182667244284872091482587763006
45	4.00000000000000000000000032098727192764668035300389656328035
46	4.00000000000000000000000008211539679679566213144479582280809
47	4.00000000000000000000000002099599390291991354365965730507100
48	4.00000000000000000000000000536578465166022788861452198401138
49	4.00000000000000000000000000137064270689761934782854129186472
50	4.00000000000000000000000000034995981272004543087585982715063
51	4.00000000000000000000000000008931473717892150619864643204065
52	4.00000000000000000000000000002278488029445791366958197682504

53	4.00000000000000000000000000000581026907354636269737558641106
54	4.00000000000000000000000000000148107951836956174433891965452
55	4.00000000000000000000000000000037739794208813320358348567379
56	4.00000000000000000000000000000009613150114596899277056034533
57	4.00000000000000000000000000000002447837919247617116131216160
58	4.00000000000000000000000000000000623097077461502353249631874
59	4.00000000000000000000000000000000158558668777775106866604971
60	4.00000000000000000000000000000000040335767047543656355201610
61	4.00000000000000000000000000000000010257966725160883998437994
62	4.00000000000000000000000000000000002607997922108963477018896
63	4.00000000000000000000000000000000000662876040731926488607073
64	4.00000000000000000000000000000000000168438150234153026989855
65	4.00000000000000000000000000000000000042789322571331107956985
66	4.00000000000000000000000000000000000010867276896030989791626
67	4.00000000000000000000000000000000000002759305787307300648501
68	4.00000000000000000000000000000000000000700448087651713462274
69	4.00000000000000000000000000000000000000177767432119150440605
70	4.00000000000000000000000000000000000000045105710581343128910
71	4.00000000000000000000000000000000000000011442390783224661917
72	4.00000000000000000000000000000000000000002902088480278385401
73	4.00000000000000000000000000000000000000000735894816187651331
74	4.00000000000000000000000000000000000000000186566878076426577
75	4.00000000000000000000000000000000000000000047290013026485080
76	4.00000000000000000000000000000000000000000011984576633465879
77	4.00000000000000000000000000000000000000000003036662502577622
78	4.00000000000000000000000000000000000000000000769295211697193
79	4.00000000000000000000000000000000000000000000194856199437495
80	4.00000000000000000000000000000000000000000000049347148987673
81	4.00000000000000000000000000000000000000000000012495062028993
82	4.00000000000000000000000000000000000000000000003163334202766
83	4.00000000000000000000000000000000000000000000000800725724571
84	4.00000000000000000000000000000000000000000000000202654474612
85	4.00000000000000000000000000000000000000000000000051281879520
86	4.00000000000000000000000000000000000000000000000012975035097
87	4.00000000000000000000000000000000000000000000000003282400078
88	4.00000000000000000000000000000000000000000000000000830260345
89	4.00000000000000000000000000000000000000000000000000209980167
90	4.00000000000000000000000000000000000000000000000000053098812
91	4.00000000000000000000000000000000000000000000000000013425645
92	4.00000000000000000000000000000000000000000000000000003394147
93	4.00000000000000000000000000000000000000000000000000000857970
94	4.00000000000000000000000000000000000000000000000000000216851
95	4.00000000000000000000000000000000000000000000000000000054802
96	4.00000000000000000000000000000000000000000000000000000013848
97	4.00000000000000000000000000000000000000000000000000000003498
98	4.00000000000000000000000000000000000000000000000000000000883
99	4.00000000000000000000000000000000000000000000000000000000223
100	4.00000000000000000000000000000000000000000000000000000000056
101	4.00000000000000000000000000000000000000000000000000000000014

Table 2

Cycle number versus the ratio of successive head-to-center distances for the midgets in our sequence

n	$\dfrac{c_n - h_n}{c_{n+1} - h_{n+1}}$
3	18.3401551994738078932293364270477004897785651570168500752764
4	16.9161183378306161702891764725654156643446147792235057394940
5	16.3085803740513327005257203105555287743134320449526969582979
6	16.0978321921328113285899928603600429128825482691063137539968
7	16.0297704370594686295146318159185441569131503544733273560405
8	16.0087874423429305973593640970435509561252722245350994806777
9	16.0025347635074281843881918406858425056673654910371699614997
10	16.0007183225733905134024956364052860491842074579038390280277
11	16.0002007520316907404888446214546727496170948444828839394015
12	16.0000554819652667942813475626919991804253118466696285494226
13	16.0000151940691096469587282777717463601460537372365258032764
14	16.0000041294185874686103949264713484271893166522228129959041
15	16.0000011150804954166347568476884785217414081172023977429657
16	16.0000002994516248697882485964200301170907943279976569103495
17	16.0000000800332844320451599356735908558664204136049505306140
18	16.0000000213009183023437288982263750885678201123057289530762
19	16.0000000056484229525721835591242374636811579891077268598887
20	16.0000000014928337032418059687293769574708956441984829448625
21	16.0000000003934171913107017701898556874707084164002808234904
22	16.0000000001034034963382996876636889847766229992526222249331
23	16.0000000000271131737125980748796832226074433941310049146265
24	16.0000000000070938683351807493790564906693819166998015290823
25	16.0000000000018523608105547634787376943709278194873666851535
26	16.0000000000004828136343287401844607767604061854160531723825
27	16.0000000000001256342665072699371036302017294626013444377153
28	16.0000000000000326412811109556568337534673224741868084741582
29	16.0000000000000084684993763806587856528094589475056406989989
30	16.0000000000000021941719182935193477140650342670851609427130
31	16.0000000000000005678004832128563027938017789791674207747931
32	16.0000000000000001467661680832563870739638702954127058537484
33	16.0000000000000000378953647558028740223109014921345458822604
34	16.0000000000000002372810458549583308209820628681165837127022
35	15.9999999999999997750126891568508622559412499651631196728737
36	16.0000000000000000006485442650120188546072574861921814672882
37	16.0000000000000000001668384988115546260638736594985423754484
38	16.0000000000000000000428852328425261346200479239420440674917
39	16.0000000000000000000110152102455409031813621972058072093856
40	16.0000000000000000000028272780701125681849765452206305325123
41	16.0000000000000000000007251883947099776575418174011007967127
42	16.0000000000000000000001858893179729543571832640451505460481
43	16.0000000000000000000000476203843171341732841206030129637264
44	16.0000000000000000000000121921097850997452137593007705871334
45	16.0000000000000000000000031197808817605708436550712294161828
46	16.0000000000000000000000007978837273833521267276081665817865
47	16.0000000000000000000000002039572731718008808520195594240952
48	16.0000000000000000000000000521132687620016530483053665531324
49	16.0000000000000000000000000133025888469821186534635608250748
50	16.0000000000000000000000000033970195370666077403133215729370
51	16.0000000000000000000000000008667728306136001353431323212752

52	16.0000000000000000000000000002210729451897706216968282579268
53	16.0000000000000000000000000000563653194299708283250403194453
54	16.0000000000000000000000000000144263348914140737153994248636
55	16.0000000000000000000000000000036876785047875028668950009118
56	16.0000000000000000000000000000174823277264000135912059302490
57	16.0000000000000000000000000002680623584714668750651575971558
58	15.9999999999999999999999999999715592425565618707462523681616
59	16.0000000000000000000000000000000153643222290414029913551954
60	16.0000000000000000000000000000000039079061431579391931396341
61	16.0000000000000000000000000000000009936829322638819096101173
62	16.0000000000000000000000000000000002525973321845697552338315
63	16.0000000000000000000000000000000000641934828257922582662834
64	16.0000000000000000000000000000000000163094081513605194310272
65	16.0000000000000000000000000000000000041426113990682435738710
66	16.0000000000000000000000000000000000010519676900740893224965
67	16.0000000000000000000000000000000000002670706325952794379190
68	16.0000000000000000000000000000000000000677873356680091362054
69	16.0000000000000000000000000000000000000172017532967996063034
70	16.0000000000000000000000000000000000000043641681691492276317
71	16.0000000000000000000000000000000000000011069745035246401051
72	16.0000000000000000000000000000000000000002807267411904931388
73	16.0000000000000000000000000000000000000000711774641249565628
74	16.0000000000000000000000000000000000000000180433107380724608
75	16.0000000000000000000000000000000000000000045730638612264526
76	16.0000000000000000000000000000000000000000011588250094837307
77	16.0000000000000000000000000000000000000000002935960134168339
78	16.0000000000000000000000000000000000000000000743714436660766
79	16.0000000000000000000000000000000000000000000188359713877967
80	16.0000000000000000000000000000000000000000000047697796781486
81	16.0000000000000000000000000000000000000000000012078562400568
82	16.0000000000000000000000000000000000000000000003101130220790
83	16.0000000000000000000000000000000000000000000001276501434379
84	16.0000000000000000000000000000000000000000000007942675591695
85	16.0000000000000000000000000000000000000000000054464061200197
86	16.0000000000000000000000000000000000000000001646024960717090
87	15.9999999999999999999999999999999999999999998291280292468739
88	16.0000000000000000000000000000000000000000000000000830371180
89	16.0000000000000000000000000000000000000000000000001031147495
90	16.0000000000000000000000000000000000000000000000010152836877
91	16.0000000000000000000000000000000000000000000000182751063798
92	16.0000000000000000000000000000000000000000000002111790070561
93	16.0000000000000000000000000000000000000000000064978156017275
94	16.0000000000000000000000000000000000000000000582204277914789
95	16.0000000000000000000000000000000000000000008649892129019734
96	15.9999999999999999999999999999999999999999999700619955097304
97	16.0000000000000000000000000000000000000000000000000000003377
98	16.0000000000000000000000000000000000000000000000000000000853
99	16.0000000000000000000000000000000000000000000000000000000215
100	16.0000000000000000000000000000000000000000000000000000000054
101	16.0000000000000000000000000000000000000000000000000000000013

9. THE PROGRAMS

The Mandelbrot generation programs written by AR are written in ANSI C and are compatible with VAX C and TURBO C. They use extended precision arithmetic routines that manipulate numbers as strings, thus giving us unlimited precision. The program iterates the function $Z = Z^2 + C$ until $|Z| > 2$ or the user-defined maximum dwell. The number of iterations for each point in the plane is saved in an ASCII file which can be decoded by the viewing programs. The program, HIMBROT, calculates the high magnification midgets and the program HIPLOT plots the image on the screen, allowing you to change the width of the dwell bands.

The cycle finder program [FINDCYC] is written in TURBO C and also uses the extended precision math routines. The cycle finder works on the principle that at a stable cycle-n in the Mandelbrot set can be found a root of a 2N degree equation.

9.1 Locating a midget

The process we used for locating a midget along the Spike evolved as we worked on this project. At the start we proceeded as follows:

1. Estimate the next major midget position based on the current major midget position using KWP's midget locator formula. One takes the current real value and differences it with -2.0. This difference is divided by 4 and the new real value is found by subtracting this new value from -2.0.

2. Refine the estimate of the midget's position using the cycle locator.

3. Find the center of radical 2 (the head of the midget) on the midget by using the cycle locator. The center of the head is the closest stable cycle-n whose value is 2 times that of the midget's.

4. Compute the magnification of the midget. The magnification is found by 1/(2.5 * (head-center - cardioid-center)) and represents the reciprocal of one half the distance from y_{min} to y_{max}.

5. Create an input file for the Mandelbrot generation program containing the center X position, center Y position, magnification, number of significant digits to use in computation, image file name, image statistics file name, x width in pixels, y width in pixels and the dwell.

6. Generate image using the program HIMBROT.

When the cycle finder program was refined we could just enter the cycle number of the major midget that we wanted to plot and the real value of its position was calculated and then entered into HIMBROT for the calculation of the picture file.

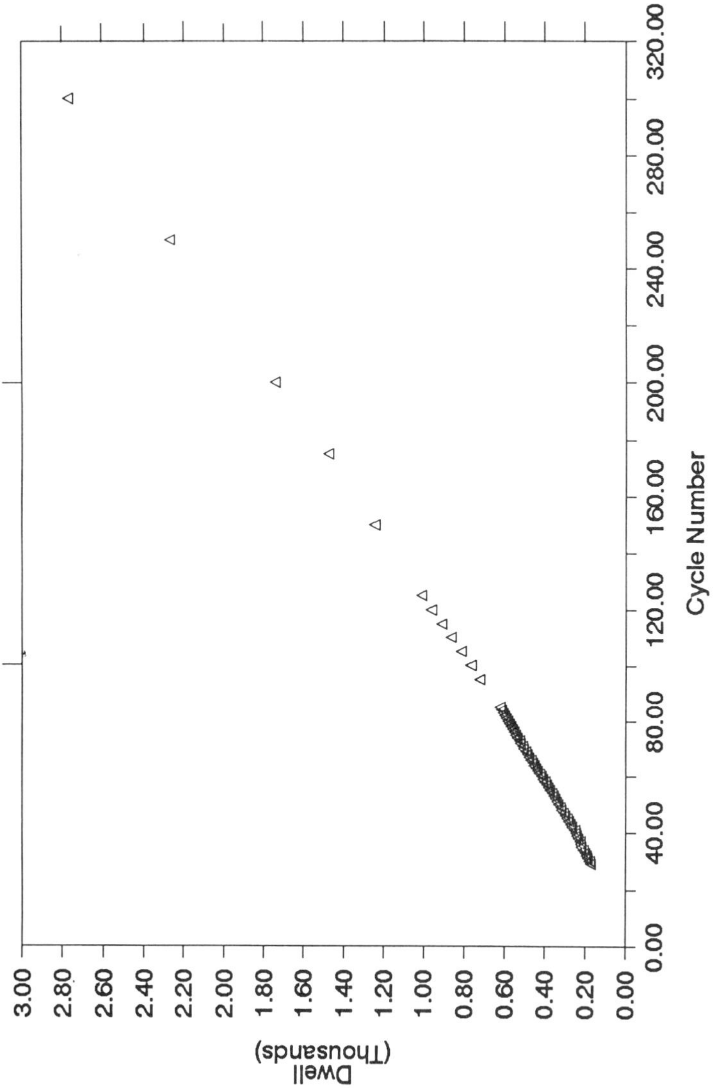

Fig. 21. Average Dwell (top of midget frames) by cycle number.

9.2 Other Programs

9.2.1 MandelMovie

MandelMovie, by Michael Larsen, Dynamic Software (distributed by Aeriel Press, Inc.) is a Macintosh II Mandelbrot/Julia set program with two unique features. It can combine a series of images into a "movie" which can be played back rapidly, and it has a fixed-point math package that allows it to work at very high magnifications (and high speed as well). The highest magnification used to date is 10^{70} diameters, but the manual states that the math precision should extend to 150 decimal places.

Larsen has sent KWP and AGDP beta versions of MandelMovie; the latest version now allows one to enter 108 decimal places in the input text boxes, the image can be computed with linear or n^2 color-to-dwell mapping, and the magnification can be run up to at least 10^{70} diameters. MandelMovie starts out plotting an image with 16x16-pixel boxes, then reduces the boxes to 8x8 pixels and re-plots, and so on until the image has been plotted at 1x1-pixel resolution. When hunting for midgets far out the Spike, one can often zero in on the coordinates one needs by the middle of the 16x16 or 8x8 pass, without having to wait the much longer time to complete the final 1x1 image. MandelMovie lacks the powerful color-editing features of a program like Platt's MandelZot, (See review by KWP (1988)) but for exploring the M-set at magnifications above the 10^{16} diameter limit of MandelZot, MandelMovie has the potentiality to be a very useful tool indeed.

9.2.2 HIMBROT and HIPLOT

The program that generates the highly magnified midgets is called HIMBROT. It is written in the C language and has been successfully compiled and run under the MS-DOS, VMS and UNIX operating systems. The program utilizes special "string arithmetic" functions and is able to achieve virtually unlimited decimal precision. The user supplies to the program, via an input data file, the number of significant numeric digits to use in computation, the width in pixels in the X direction, the width in pixels in the Y direction, the maximum number of iterations to perform on a point in the complex plane, the center co-ordinates of the desired midget and the magnification of the midget. From the supplied center and magnification values the location of the four corners of the image are computed. The program then moves through the "box" created by the four corners and iterates the function $F_c(Z) = Z^2 + C$ on each corner point until $|Z| > 2$ or until the maximum dwell has been reached. The program keeps track of how many iterations it took to complete each point and writes this value to a data file, one number per line. Because of the great amount of time it takes to complete a picture, HIMBROT has the ability to be stopped at any point in its generation of a midget and be re-started without loosing more than the calculation time spent on the last pixel being iterated.

The cycle finder finds a real root of the equation using the secant method. The program is supplied with two initial guesses of where the root is located, the cycle number of the midget to be found, and an epsilon representing the accuracy of the final approximation. The program then iterates the function $F(z) = Z^2 + C$ (where Z and C are real) N times where N is the value of the cycle number. By iterating the function N times it is in fact evaluating a function of degree 2^N. The program keeps iterating the function and refining the estimated root until the difference between the new guess and the old guess is less than epsilon. The final value output by the program is an approximation of a root of the equation and corresponds to a stable cycle-n in (**M**).

Because of the symmetry of the Mandelbrot set along the real axis HIMBROT only calculates the part of the midget that resides in the positive imaginary half-plane, thus reducing the necessary calculation time by 50%. The viewing program HIPLOT reads the data file created by HIMBROT and displays the image reflected about the real axis, thus producing the desired image.

9.2.3 FINDCYC

Our sequence of midgets contains one cycle-n midget for each n > 2. The center, c_n, of our n-cycle midget is the smallest real solution of the polynomial equation $f_n(c) = 0$, where this polynomial is defined inductively by

$$f_1(c) = c, \text{ and for } n > 0,\ f_{n+1}(c) = (f_n(c))^2 + c.$$

(See Chapter 4.) To find c_n, the Secant Method is applied. Recall this method starts with two guesses, g_1 and g_2, for the solution and produces a third guess, g_3. This point g_3 occurs where the straight line through the points $(g_1,f_n(g_1))$ and $(g_2,f_n(g_2))$ intersects the c-axis. Then g_2 and g_3 are used to produce g_4, g_3 and g_4 to produce g_5, and so on. From established properties of the polynomials $f_n(c)$, we know starting from $g_1 = -2 -e$ and $g_2 = -2$ (e is a small positive number), the Secant Method produces a sequence of guesses converging to c_n. For each application of $f_n(c)$, the inductive process (1) is carried out by extended precision arithmetic routines via string manipulations. The Secant Method is iterated until the difference between successive guesses drops below a user-specified tolerance.

9.2.4 V63MBROT

V63MBROT is a program written by Michael Freeman, Vancouver, Canada, specifically for the ATI VGA-wonder video card, mounted in an IBM or IBM-compatible personal computer. It can compute 256-color images in a 800x600 pixel format. There is a color editing mode which allows you to see each pixel of the image and the color of that pixel can be changed by use of three RGB color bars. All pixels of that dwell value will be changed along with the selected pixel. The cursor can be moved to another area of the picture and the dwell values between the selected pixel and the newly selected pixel will be changed in smooth steps between the two colors selected. You have control of the dwell, escape radius and can enter positions by their corner values or by entering the central position and the magnification. Images can be stopped and then restarted later. A companion program, LJDUMP, will plot the image on a LaserJet along with the data concerning the values of the parameters used in the generation of the image.

9.2.5 FRACTINT (Version 15)

FRACTINT (now in version 16) is a marvelous program written by a group of fractal enthusiasts. It is freeware and can be obtained from Compuserve, among other places. Information about this program is circulated on the computer discussion list, FRAC-L. The program runs on IBM PCs and compatibles. Of all the programs we have seen on the IBM side, it is the fastest one in calculating an image. It also has the most diverse collection of fractal images to plot. At 300 dots per inch, FRACTINT plots 5.3x3.9 cm images and there are many examples of these plots in this book. Fractint is described in a new book, **Fractal Creations** (Wegner and Peterson 1991). There is a disk included with the book that contains the Fractint program for IBM PCs and clones. There is also a set of colored glasses to view the 3D images that can be created with the program.

REFERENCES

Devaney, R. L. 1989 An Introduction to Chaotic Dynamical Systems, 2nd Edition, Addison Wesley, Menlo Park
Devaney, R. L. 1990 Chaos, Fractals, and Dynamics, Addison-Wesley, Menlo Park
Dewdney, A. K. 1985 Scientific American 253, 16 (August Issue)
Dewdney, A. K. 1989 Scientific American 260, 122 (June Issue)
Frame, M., Philip, A. G. D. and Robucci, A. 1991 Computers and Graphics, in press
Gleick, J. 1987 Chaos, Viking, New York
Mandelbrot, B. 1983 The Fractal Geometry of Nature, W. H. Freeman, New York
Milnor, J. 1989 Self-Similarity and Hairiness in the Mandelbrot Set, Computers in Geometry and Topology, M. Tangora, ed, Maracel Dekker, New York, p. 211-257
Milnor, J. 1991 Dynamics and Fractals, Winifred Asprey Lecture, Vassar College, Poughkeepsie, NY
Namreh, E. N. (Rollo Silver) 1987 Amygdala 5, 2
Peitgen, H. -O. and Richter, P. H. 1986 The Beauty of Fractals, Springer Verlag, Berlin
Peitgen, H. -O. and Saupe, D. 1988 The Science of Fractal Images, Springer-Verlag, New York
Penrose, R. 1989 The Emperor's New Mind, Oxford University Press, New York
Philip, A. G. D. and Philip, K. W. 1990 Amygdala 19, 1
Philip, A. G. D. and Robucci, A. 1991 Amygdala 23, 23
Philip, K. W. 1988 Amygdala, 14, 1
Stewart, I. 1989 Does God Play Dice?, Basil Blackwell, Worcester
Stewart, I. 1987 The Problems of Mathematics, Oxford University Press, Oxford
Wegner, T. and Peterson, M. 1991 Fractal Creations, The Waite Group Press

10. APPENDICES

APPENDIX I - THE GALLERY OF MIDGETS

APPENDIX II - MATHEMATICAL APPENDIX

APPENDIX III - TABLE OF POSITIONS

10.1 APPENDIX I - THE GALLERY OF MIDGETS

Images of the Cycle 3 - Cycle 85 midgets are shown, followed by images of Cycles 95 - 125 by intervals of 5 cycles, Cycle 125 - 200 by intervals of 25 cycles and Cycle 200 - 300 by intervals of 50 cycles. Below each image the Real position, magnification, resolution (in pixels), maximum dwell and computer system used are listed.

Following the Cycle 300 midget there are eight additional pictures of some of the midgets at higher resolution. Cycles 28, 39, 44, 49, 57 and 60 were computed by KWP using MandelMovie. Cycle 70 was computed by AGDP using MandelMovie and Cycle 85 was computed by AR using HIMBROT. It is interesting to compare these images with the corresponding 100x76 pixel images. The low resolution midgets are now presented at a resolution that shows their similarity to (**M**).

These pictures are much enlarged over the image that appears on the screen. The program PIZZAZ+ (Applications Techniques) was used to print the images on an HP LaserJet printer. On the Cycle 39 picture there is a small inset (in the lower right corner) that more nearly approximates what is seen on the screen of the computer.

Cycle 3 R = -1.7548 M = 2.78×10^{1} S = 640x480 D = 1000 AR FRA12 AT

Cycle 4 R = -1.94079 M = 4.42×10^2 S = 640x480 D = 1000 AR FRA12 AT

Cycle 5 R = -1.985424 M = 8.62×10^3 S = 640x480 D = 1000 AR FRA12 AT

Cycle 6 R = -1.9963761 M = 1.40×10^5 S = 640x480 D = 1000 AR FRA12 AT

Cycle 7 R = -1.99909568 M = 2.26×10^6 S = 640x480 D = 1000 AR FRA12 AT

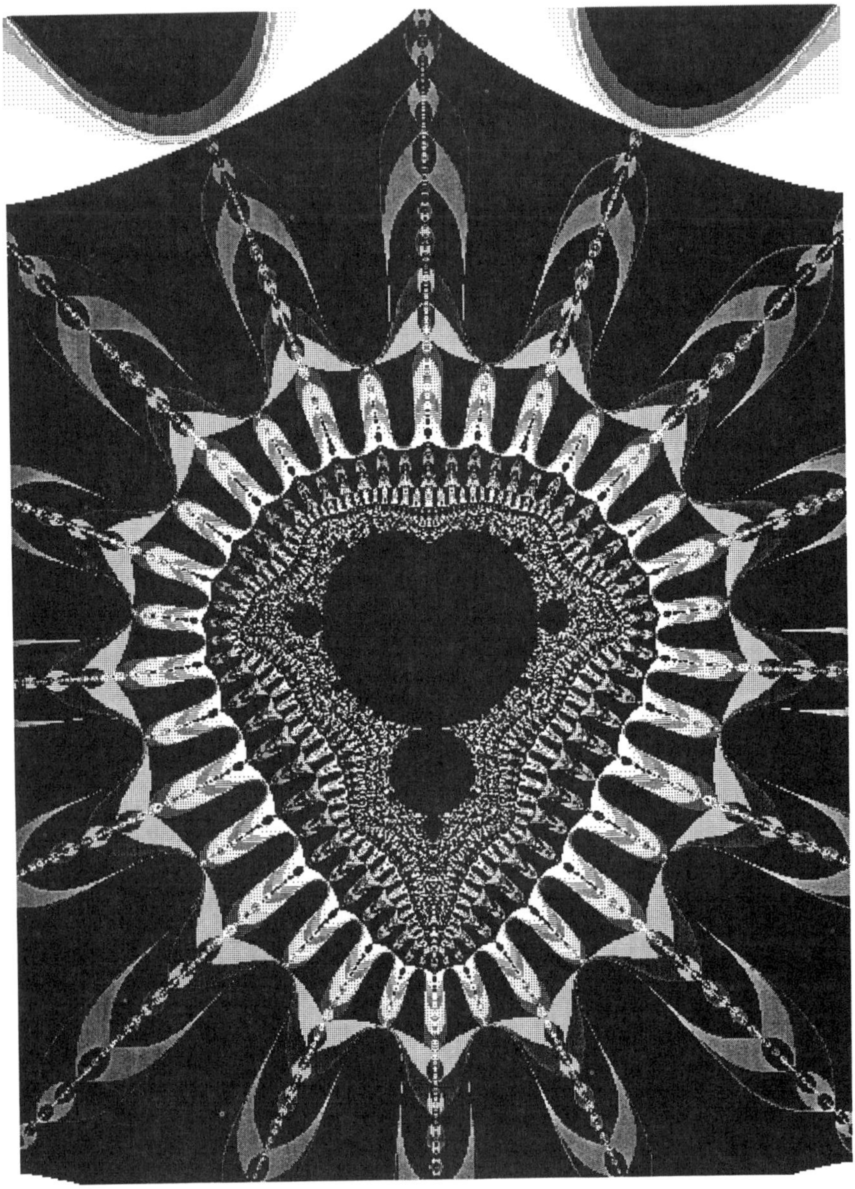

Cycle 8 R = -1.999774048 M = 3.62×10^7 S = 640x480 D = 1000 AR FRA12 AT

Cycle 9 R = -1.9999435217 M = 5.81×10^{8} S = 640x480 D = 1000 AR FRA12 AT

Cycle 10 R = -1.9999858811 M = 9.28×10^9 S = 640x480 D = 1000 AR FRA12 AT

Cycle 11 R = -1.99999647033350 M = 1.49×10^{11} S = 640x480 D = 1000 AR FRA12 AT

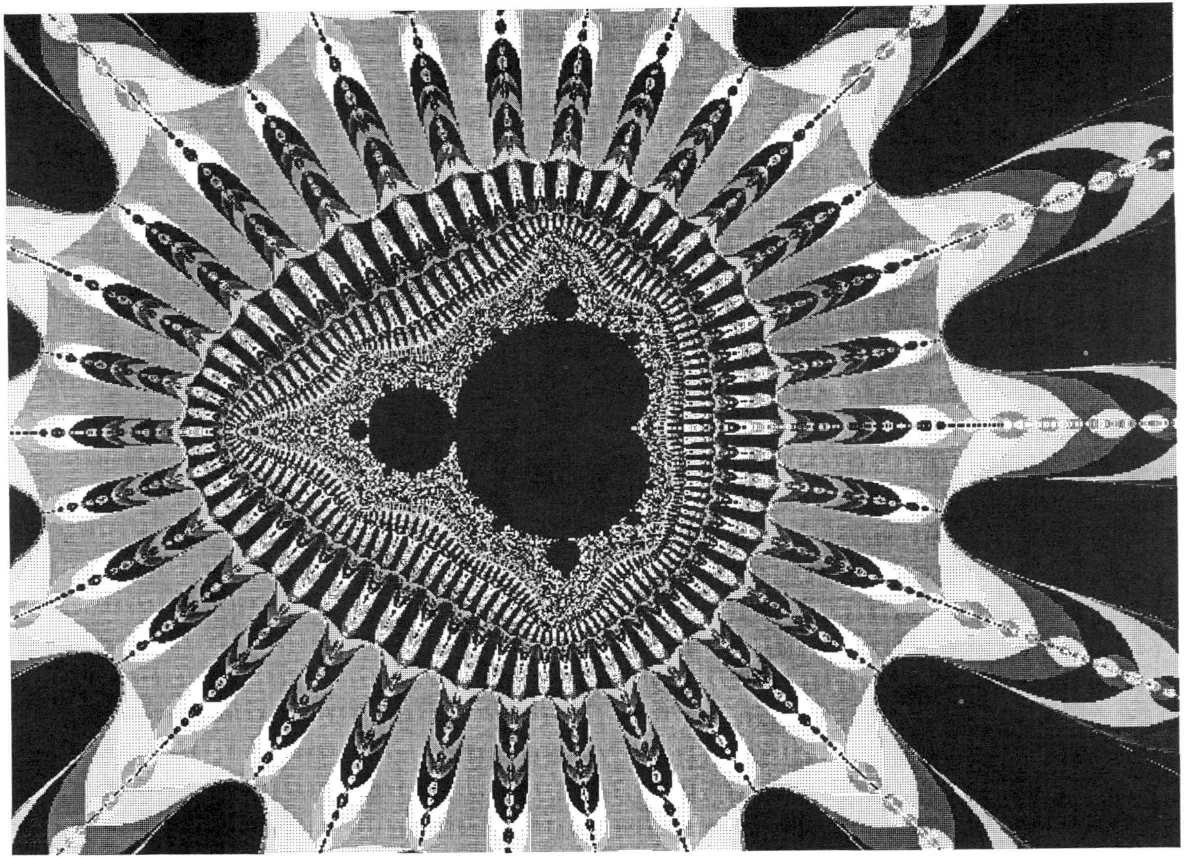

Cycle 12 R = -1.999999117587260 M = 2.35x10^{12} S = 640x480 D = 1000 AR FRA12 AT

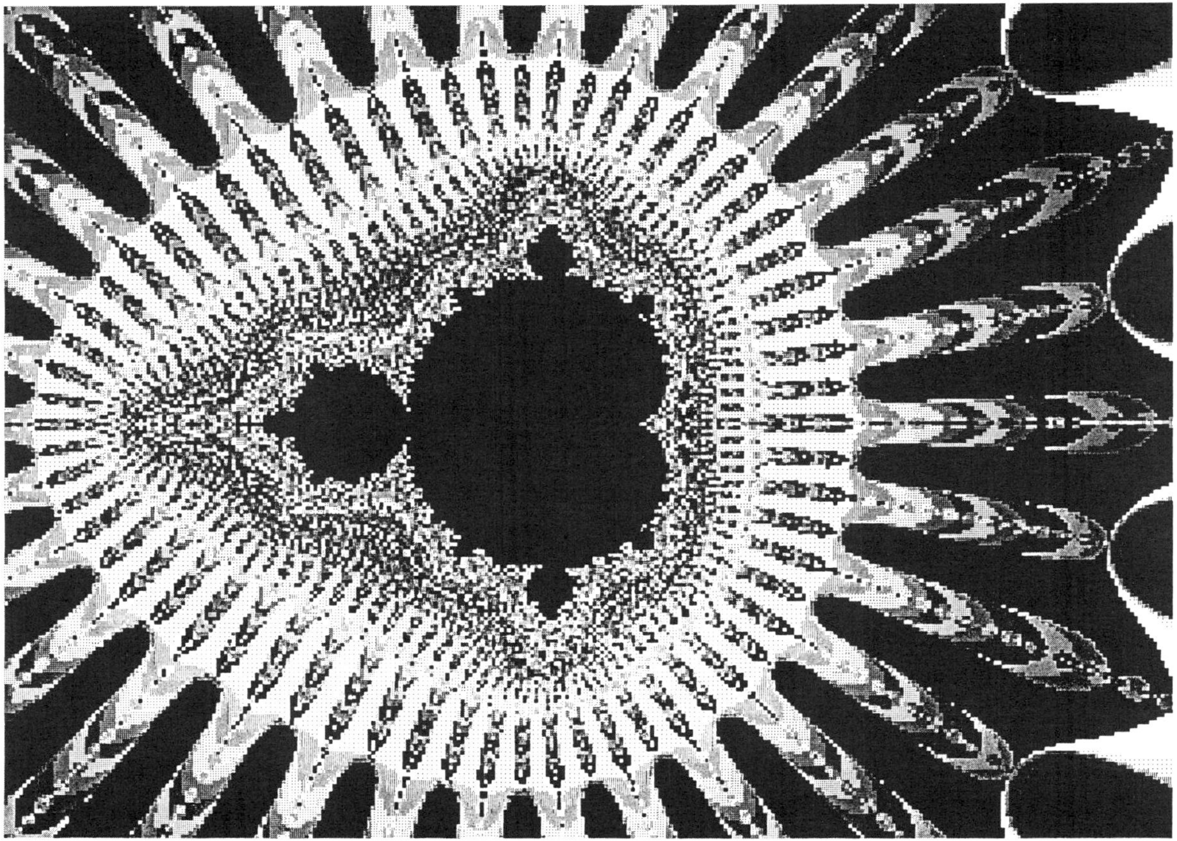

Cycle 13 R = -1.9999997793970588 M = $3.04x10^{13}$ S = 400x300 D = 1500 AR VAX

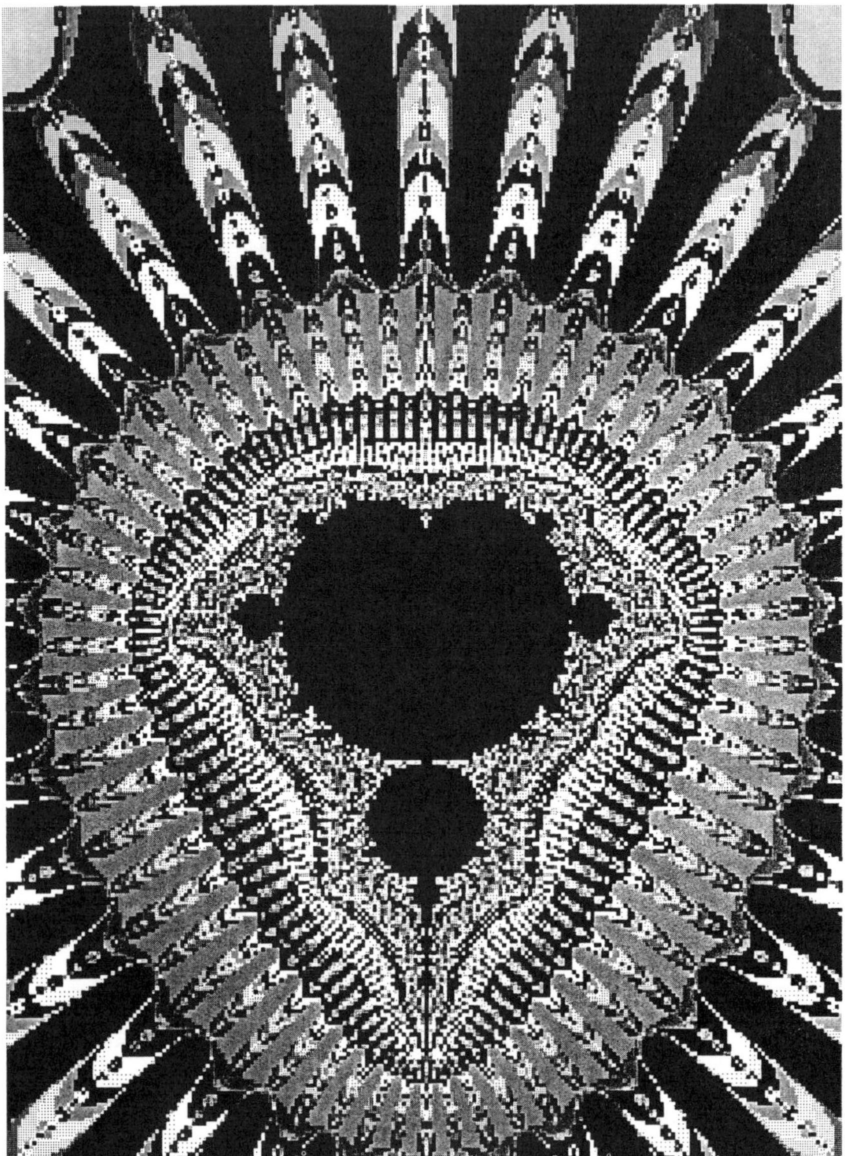

Cycle 14 R = -1.99999994484928145 M = 4.87×10^{14} S = 400x300 D = 1500 AR VAX

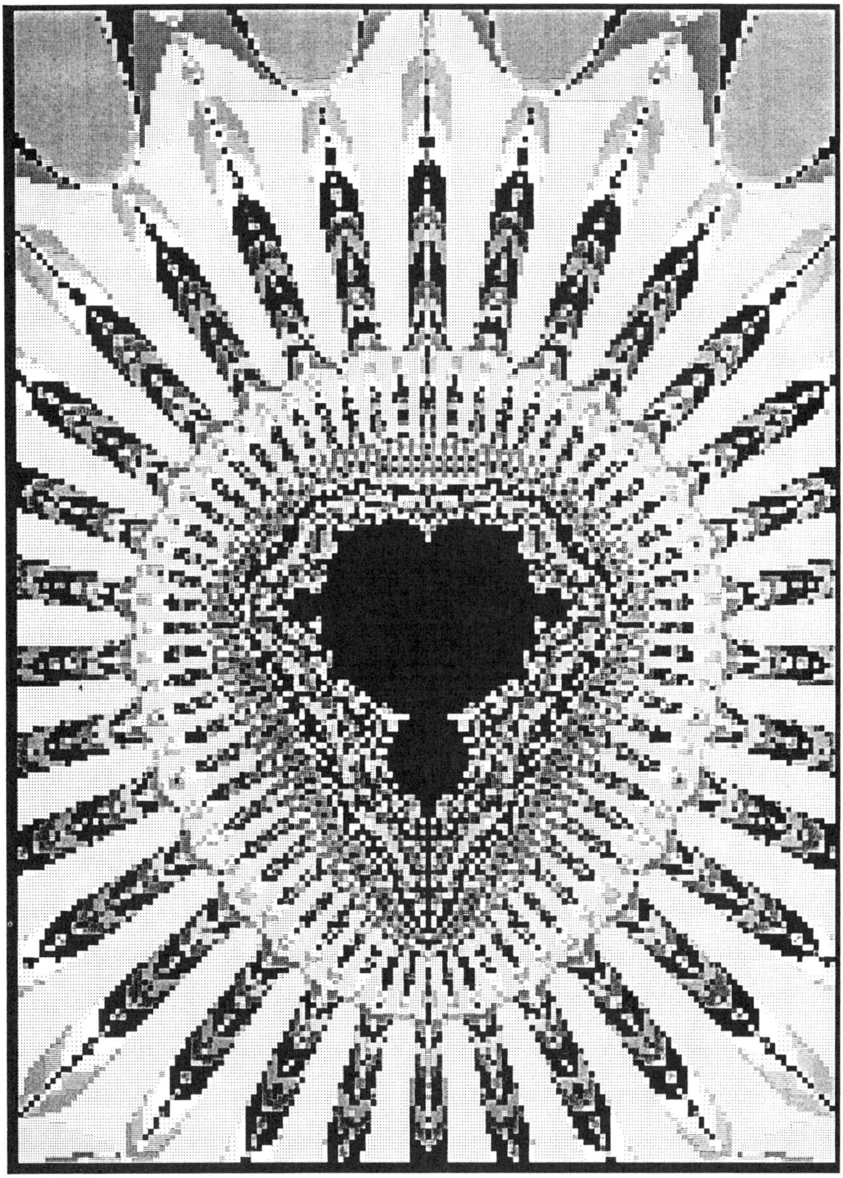

Cycle 15 R = -1.999999986212321505 M = 7.79×10^{15} S = 200x150 D = 1000 AR VAX

Cycle 16 R = -1.9999999965530804536 M = 1.24×10^{17} S = 200x150 D = 1000 AR VAX

Cycle 17 R = -1.99999999913827011860 M = 1.99×10^{18} S = 200x150 D = 1000 AR VAX

Cycle 18 R = -1.9999999997845675300000 M = 3.19×10^{19} S = 200x150 D = 1000 AR VAX

Cycle 19 R = -1.9999999999461418825233 M = 5.104x10[20] S = 200x150 D = 1000 AR VAX

Cycle 20 R = -1.9999999999865335470632367 M = 1.02×10^{22} S = 200x150 D = 1000 AR VAX

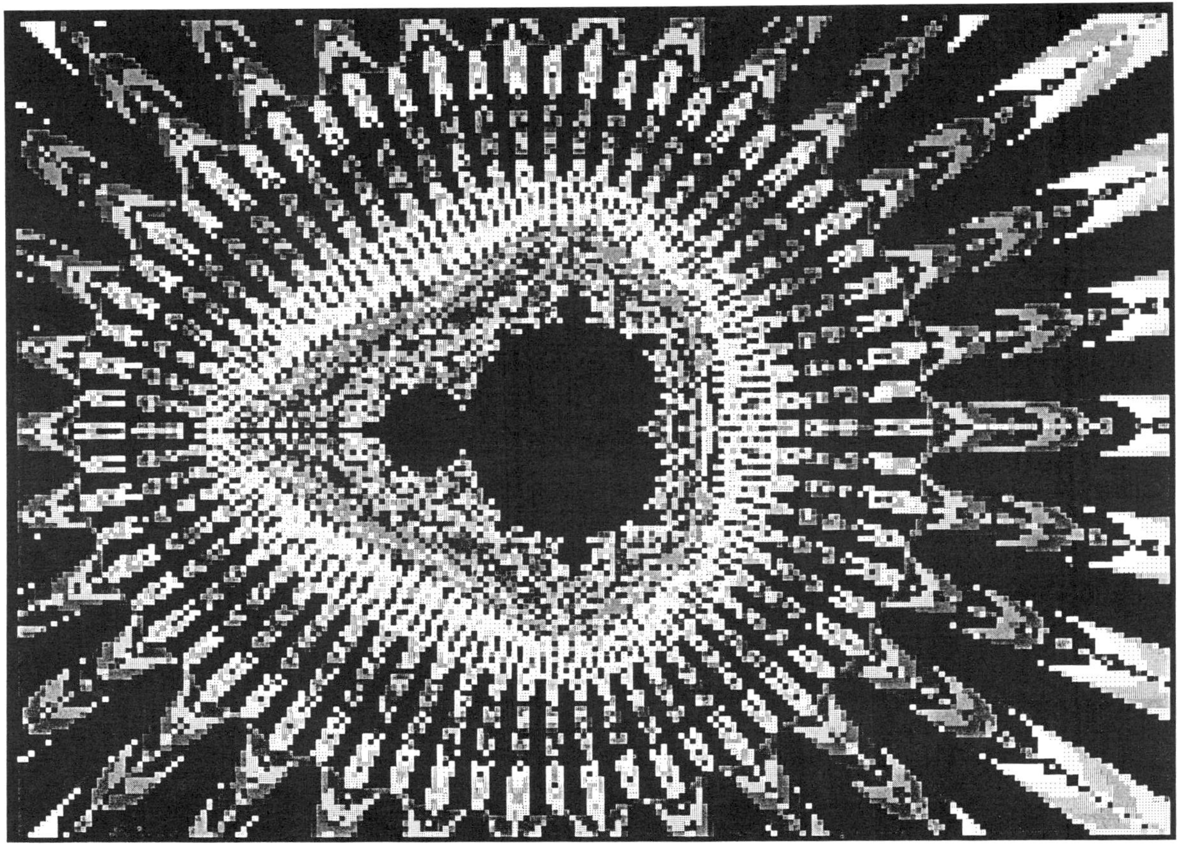

Cycle 21 R = -1.999999999999966338676581938 M = 1.31×10^{23} S = 200x150 D = 1000
AR VAX

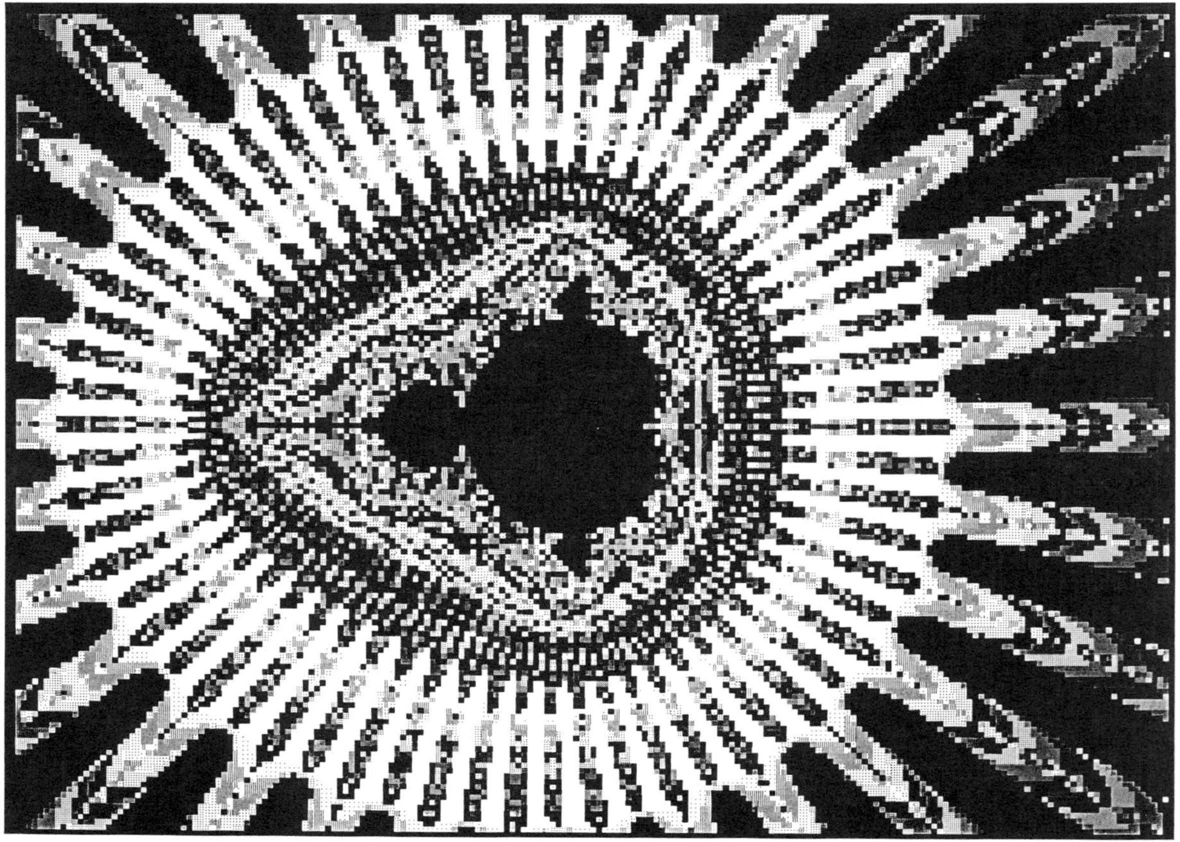

Cycle 22 R = -1.999999999999158466914555203 M = 2.09x10^{24} S = 200x150 D = 1000 AR VAX

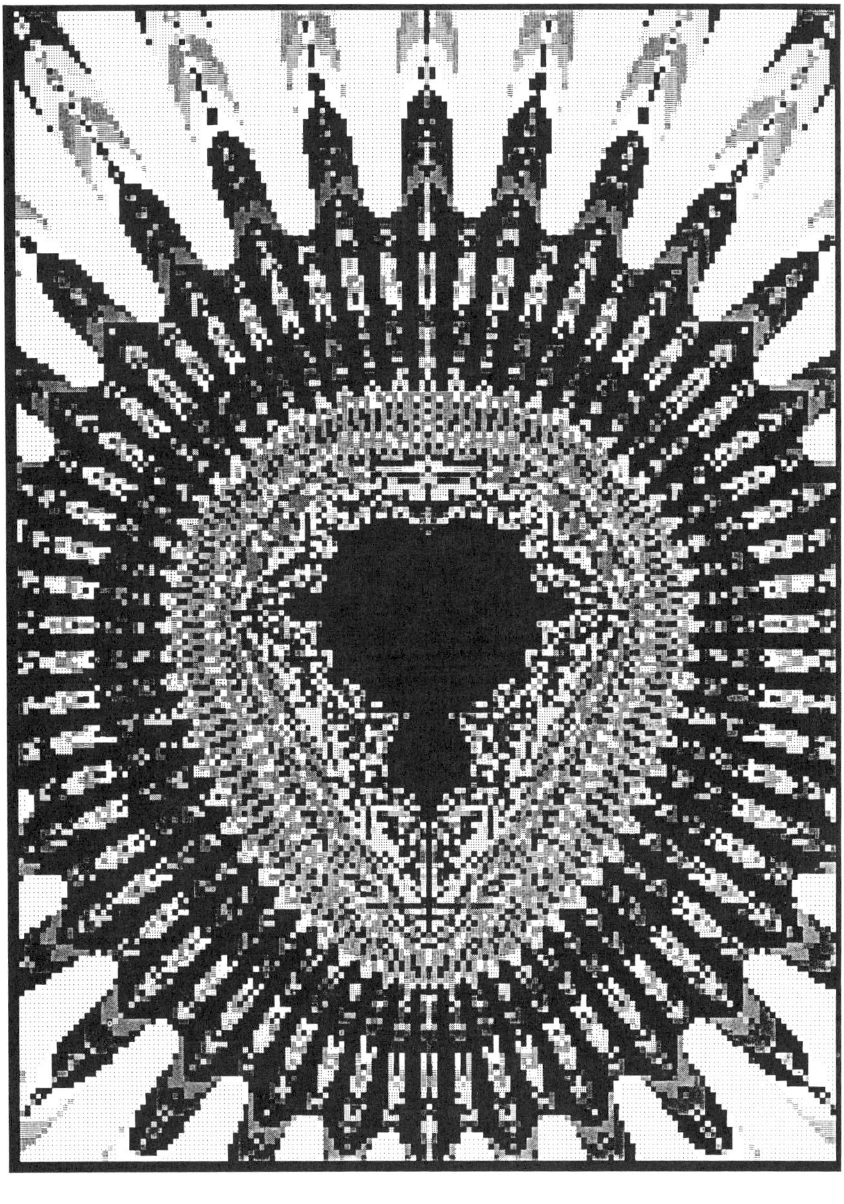

Cycle 23 R = -1.99999999999789616728639243 79 M = 3.34×10^{25} S = 200x150 D = 1000
AR VAX

Cycle 24 R = -1.99999999999999947404182159840012 M = 5.35×10^{26} S = 200x150 D = 1000
AR VAX

Cycle 25 R = -1.9999999999999868510455399619060 M = 8.56×10^{27} S = 200x150
D = 1000 AR VAX

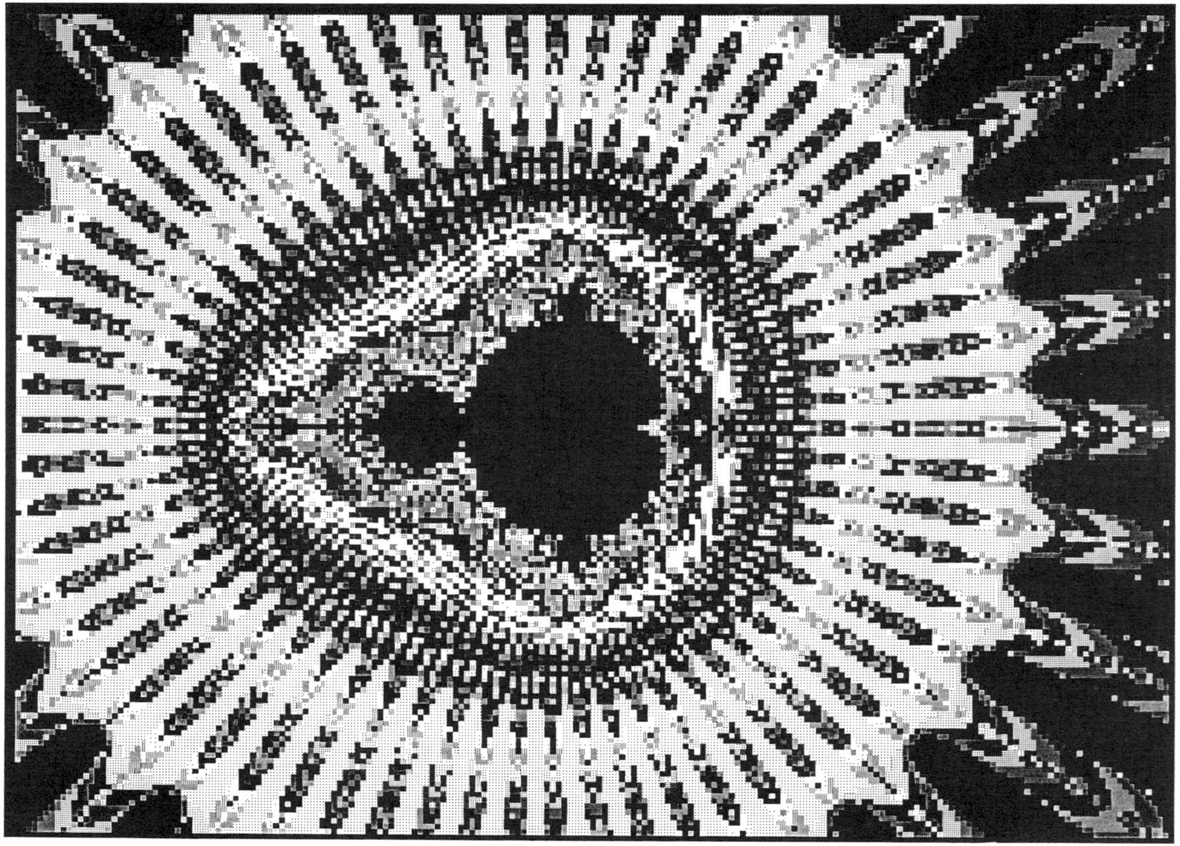

Cycle 26 R = -1.9999999999999967127613849906008 M = 1.35×10^{29} S = 200x150
D = 1000 AR VAX

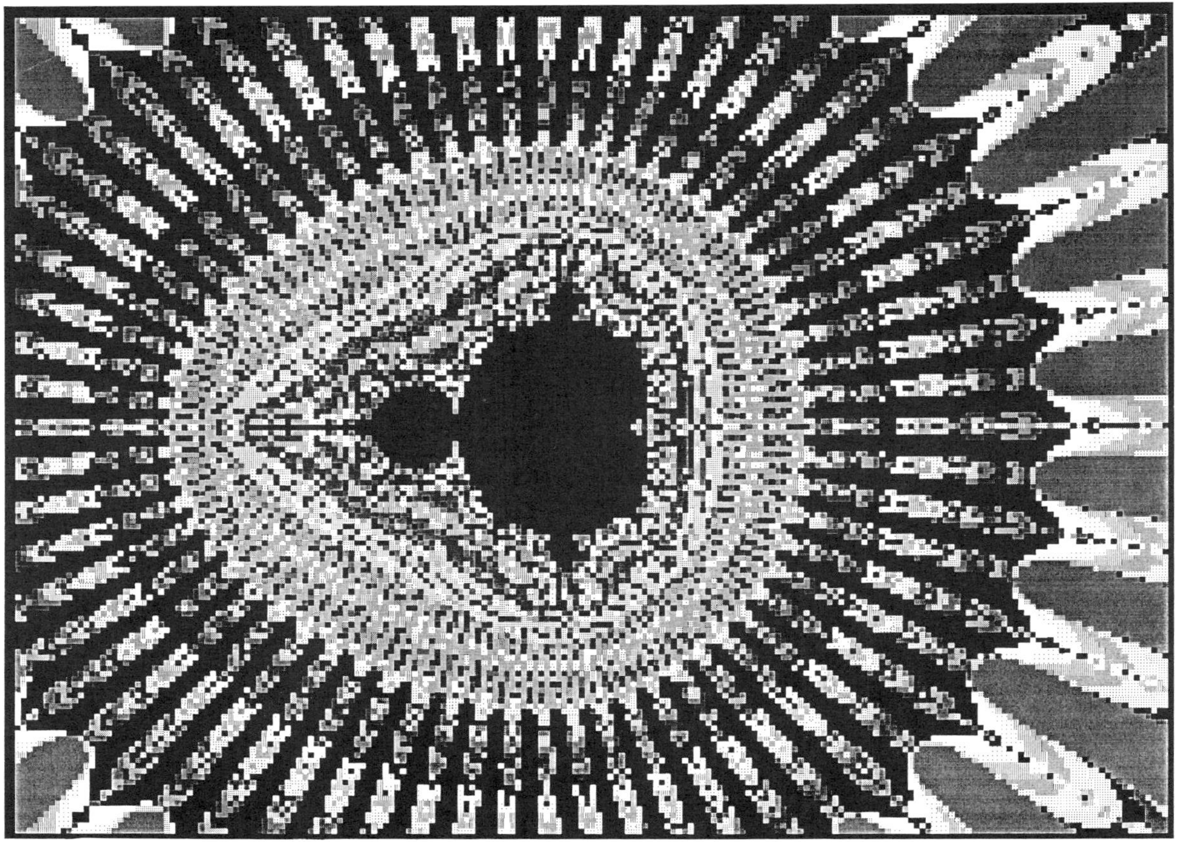

Cycle 27 R = -1.9999999999999999178190346247658319 M = 2.19×10^{30} S = 200x150
D = 1000 AR VAX

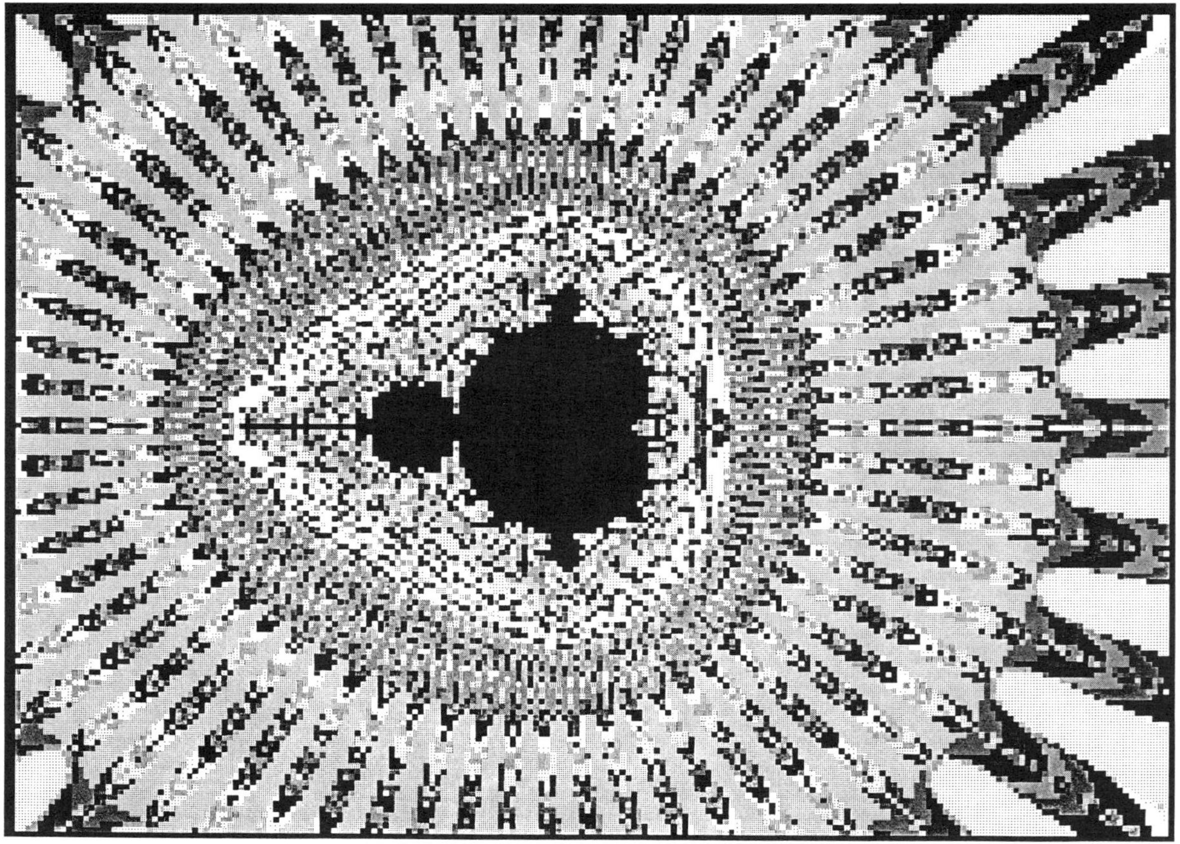

Cycle 28 R = -1.99999999999999997945475865619151078 M = 3.51×10^{31} S = 400x300
D = 1500 AR VAX

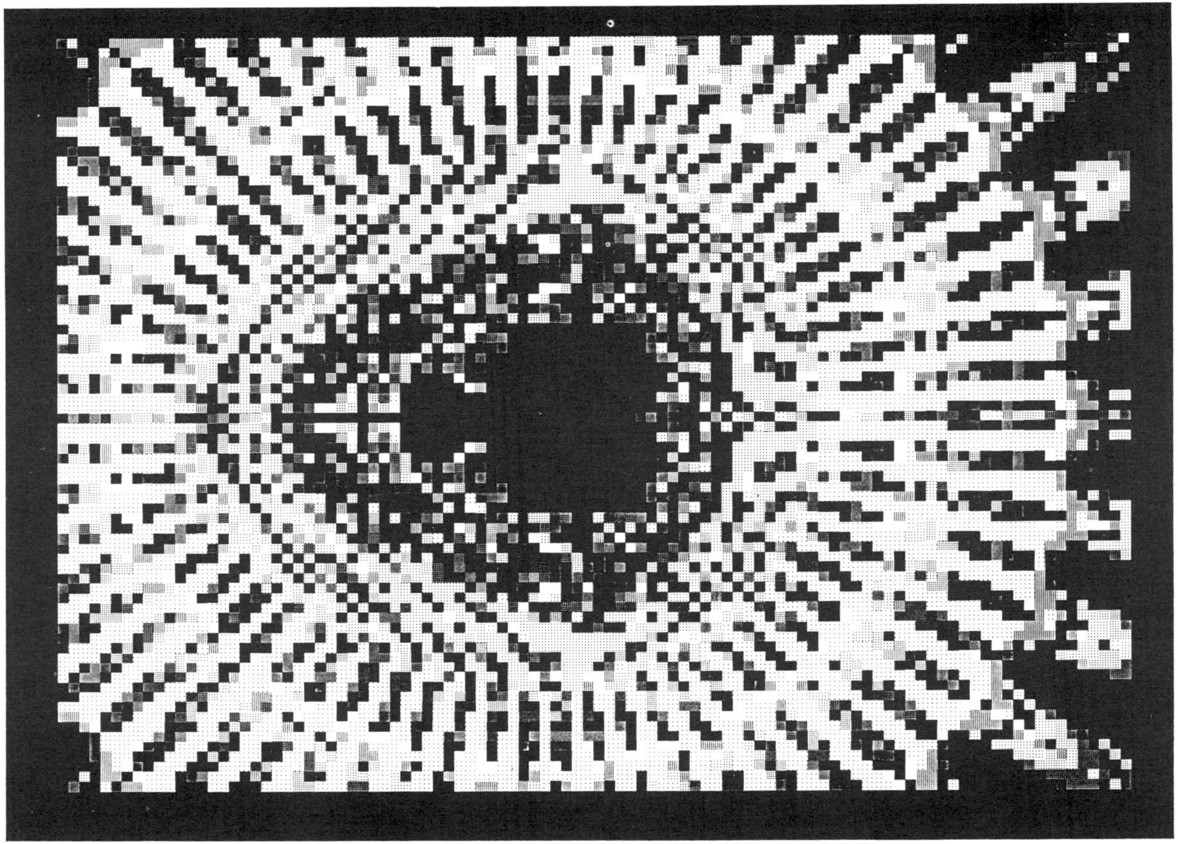

Cycle 29 R = -1.99999999999999994863689664047881127 M = 3.51×10^{32} S = 200x150
D = 1500 AR VAX

Cycle 30 R = -1.99999999999999987159224160119705045 M = $5.62x10^{33}$ $ = 200x150
D = 1500 AR VAX

Cycle 31 R = -1.9999999999967898060400299264058 M = $1.56x10^{35}$ S = 100x75
D = 1000 AR VAX

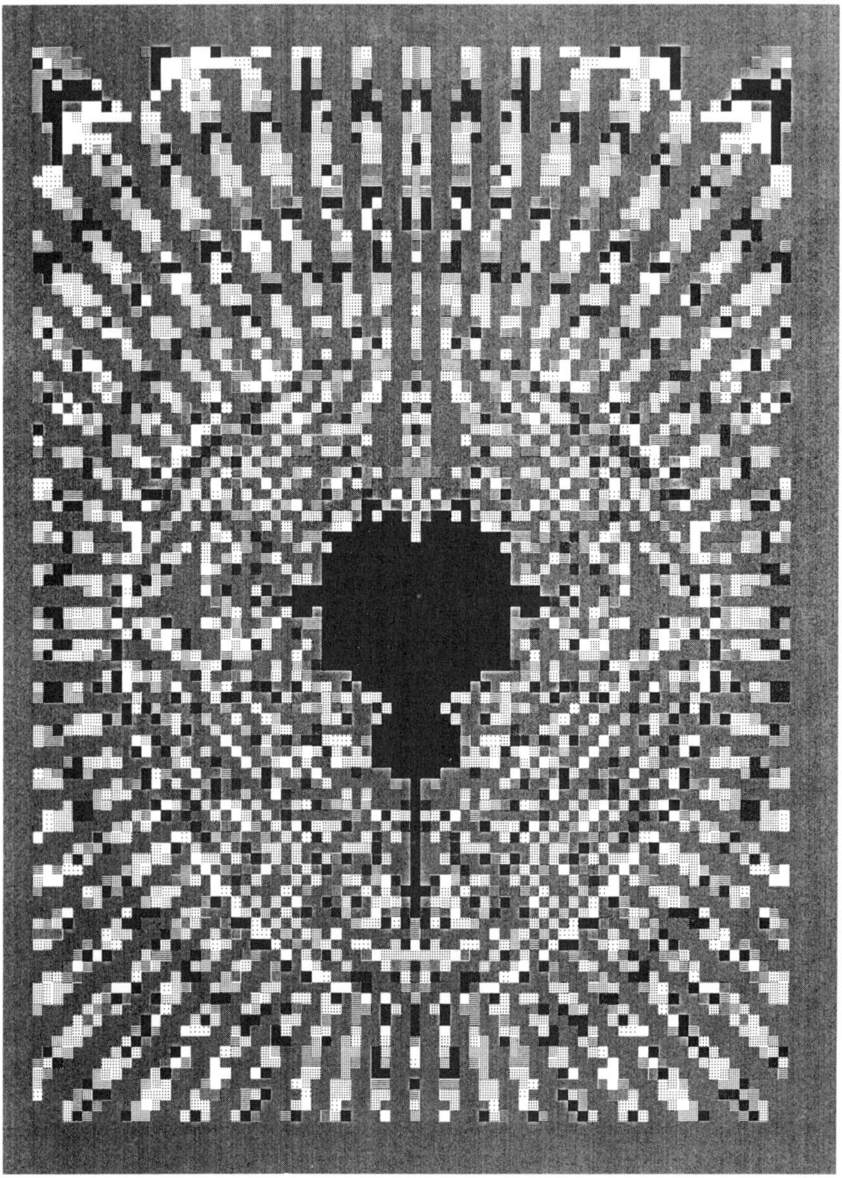

Cycle 32 R = -1.999999999919745151000748161079 M = 4.69x10^{36} S = 100x75
D = 1000 AR VAX

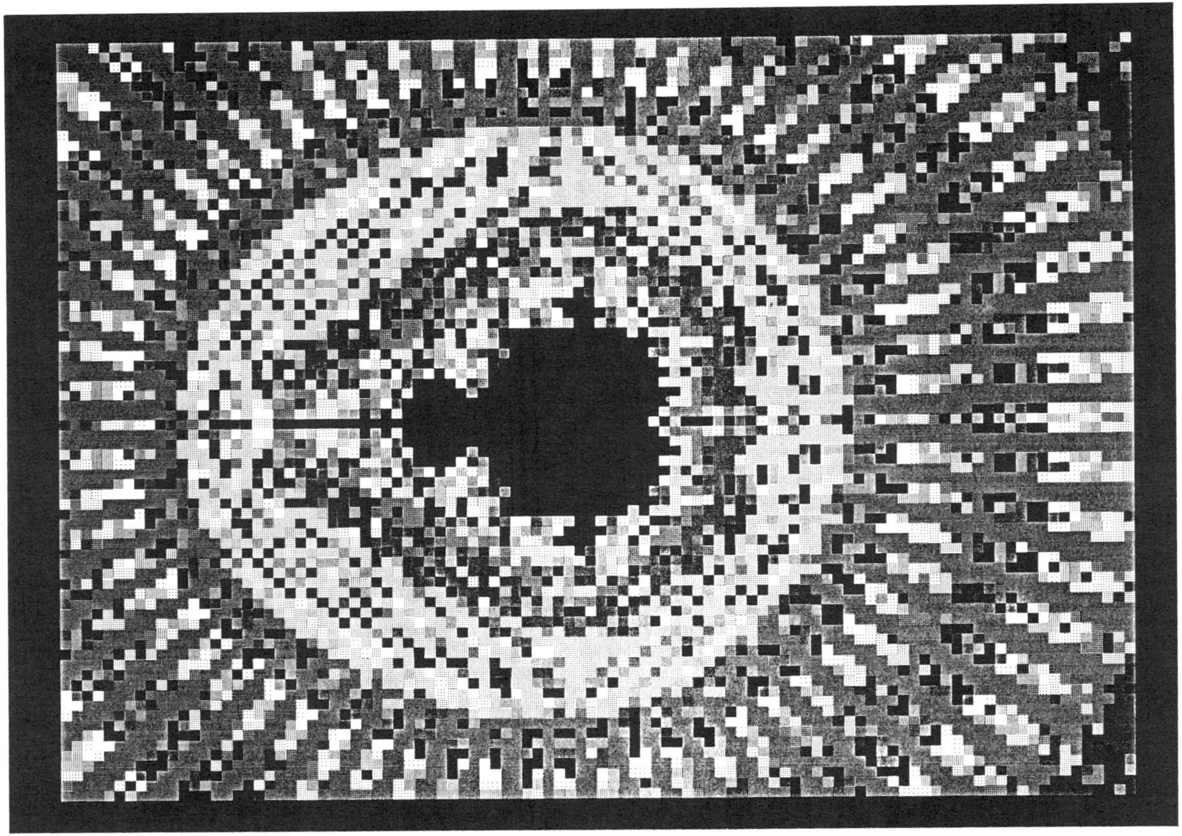

Cycle 33 R = -1.999999999999999999799362877501870403303 M = 4.60×10^{37} S = 100x75
D = 1000 AR VAX

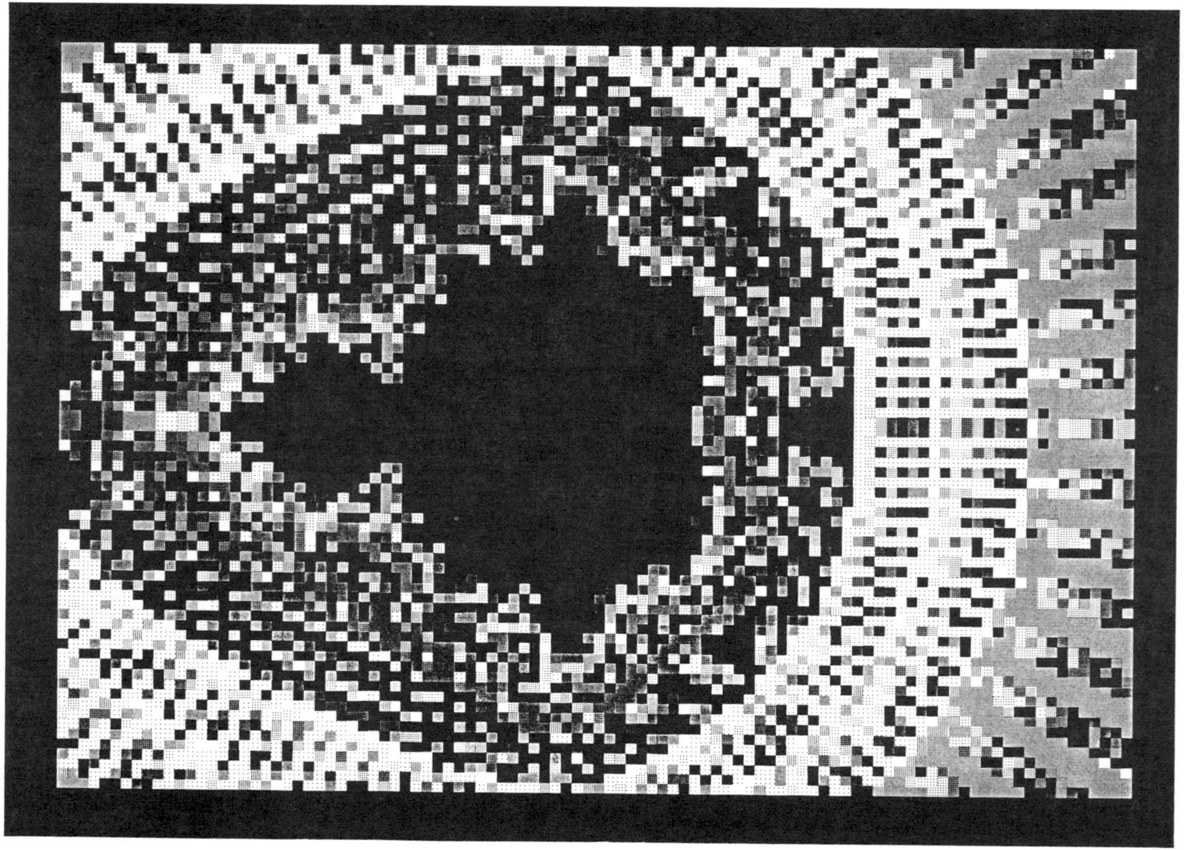

Cycle 34 R = -1.99999999999999999994984071937546760086492 M = 1.56x10^{39}
S = 100x75 D = 1000 AR VAX

Cycle 35 R = -1.99999999999999999987460179843866900218748 M = 2.00×10^{40}
S = 100x75 D = 1000 AR VAX

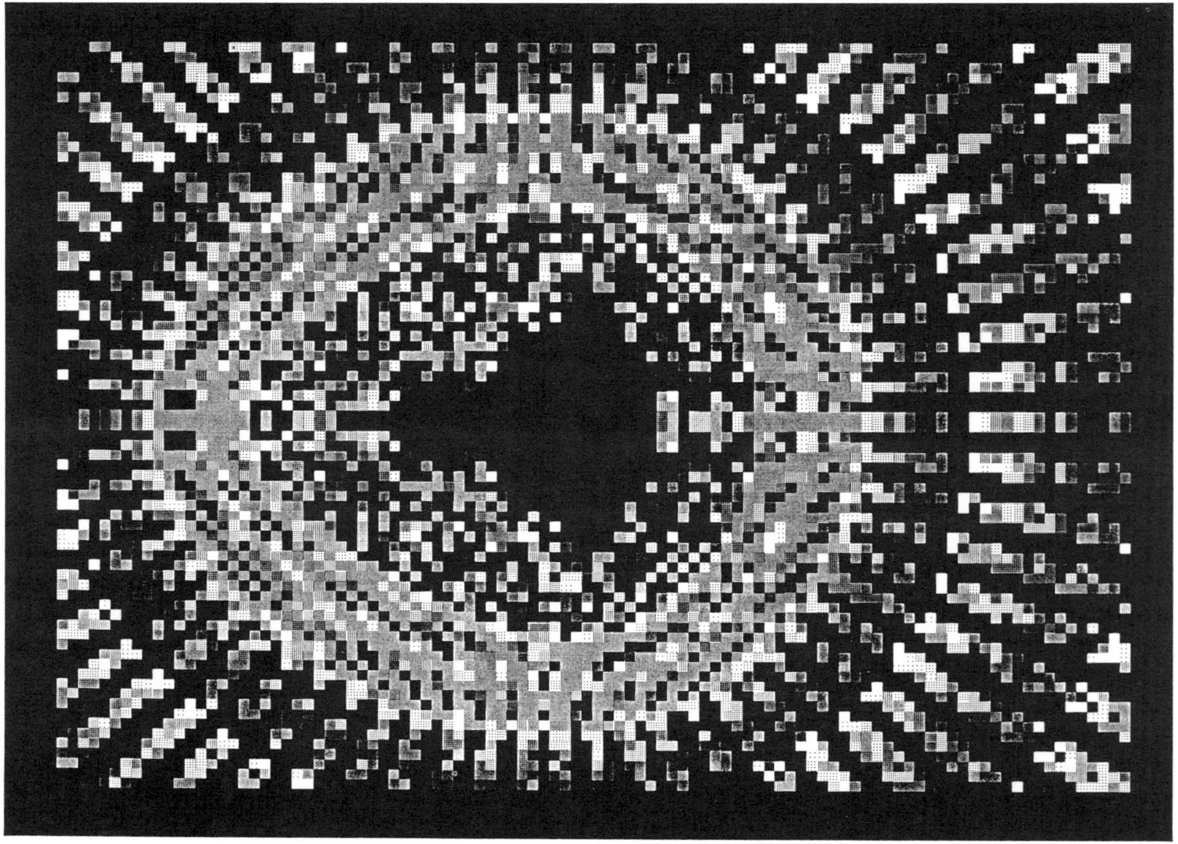

Cycle 36 R = -1.9999999999999999999968650449609667250548 4931 M = $4.30 x 10^{41}$
S = 100x75 D = 1000 AR VAX

Cycle 37 R = -1.9999999999999999999999216261240241681263722775 M = 2.50x10^42
S = 100x75 D = 1000 AR AT

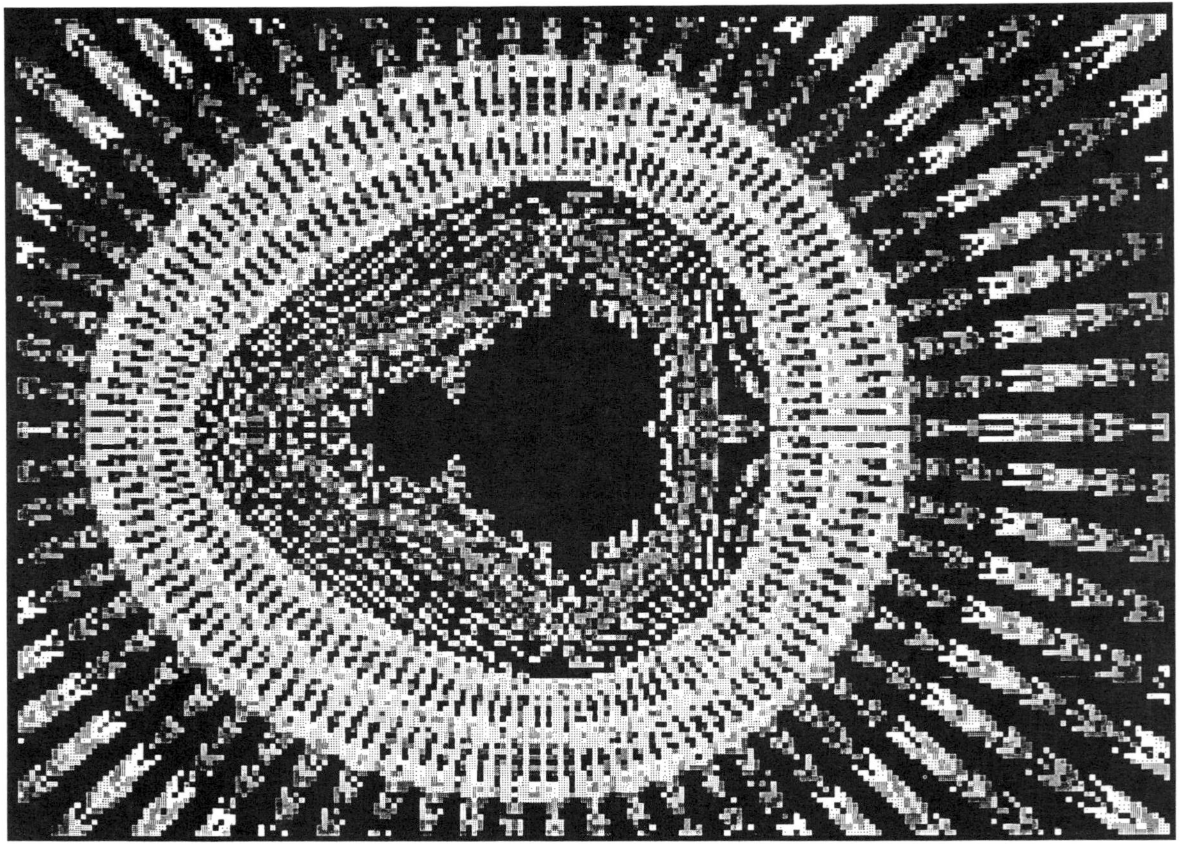

Cycle 38 R = -1.999999999999999999999980406531006042031593136 60 M = 5.88x10^{43}
S = 200x150 D = 1000 AR AT

Cycle 39 R = -1.99999999999999999999995101632751510507898288471 M = 7.06x10^{44}
S = 100x75 D = 1000 AR AT

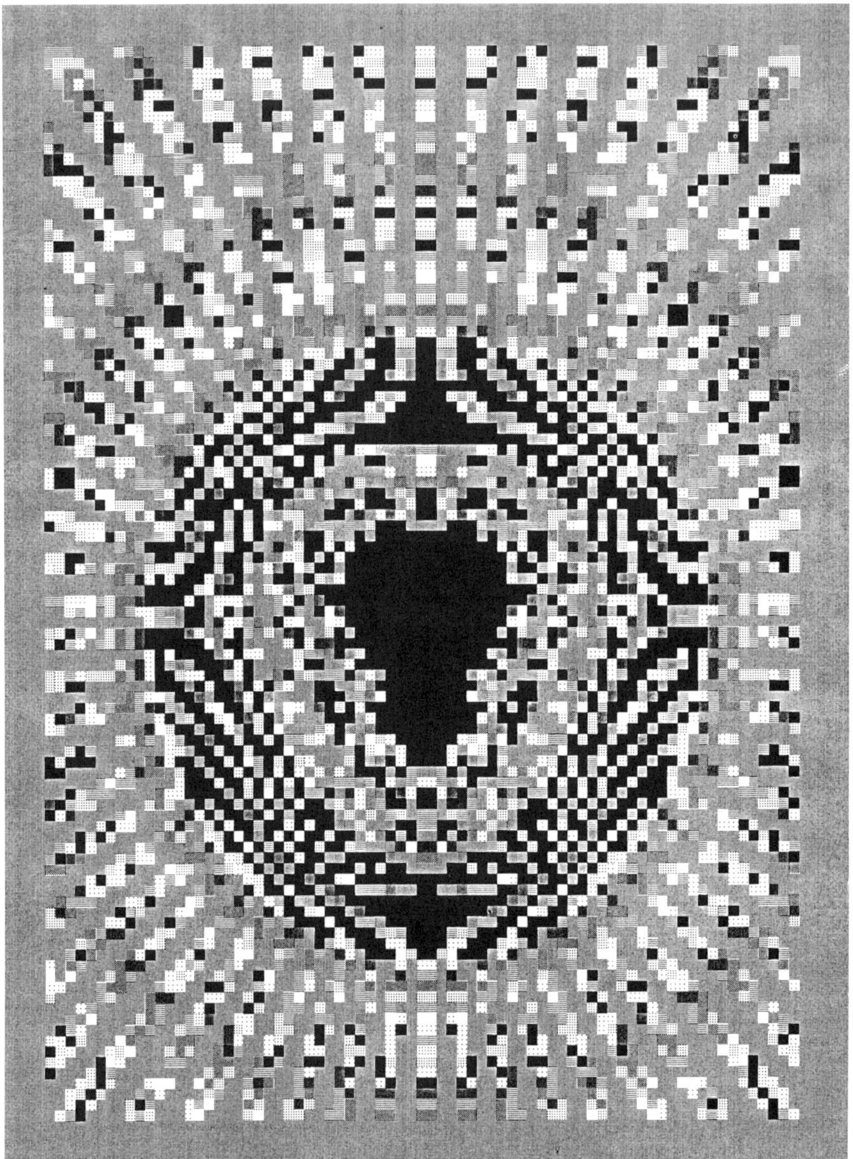

Cycle 40 R = -1.99999999999877540818787762697457239 52
M = $9.97 x 10^{45}$ S = 100x75 D = 1000 AR AT

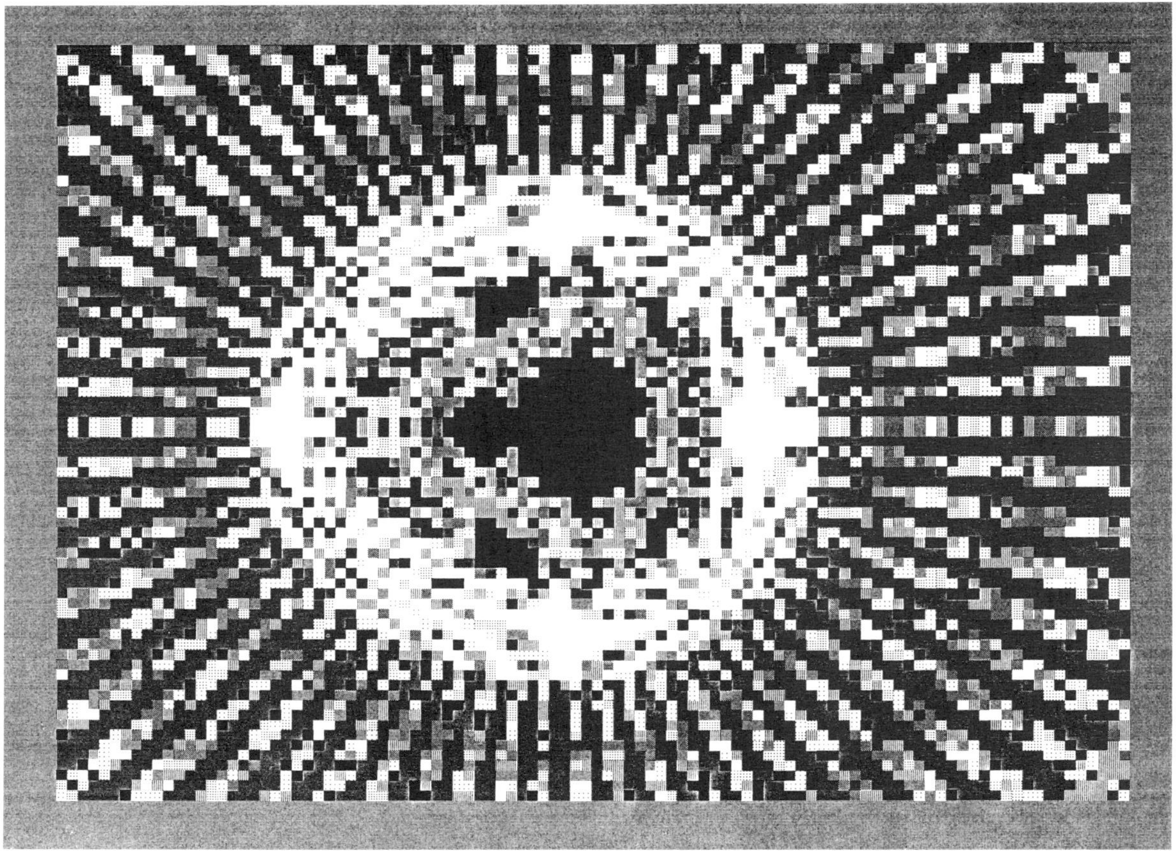

Cycle 41 R = -1.999999999999999999999969385204696940674364311663
M = 1.41×10^{47} S = 100x75 D = 1000 AGDP ZEN

Cycle 42 R = -1.99999999999999999999999234630117423516859107803 01
M = $3.16x10^{48}$ S = 100x75 D = 1000 AR AT

Cycle 43 R = -1.999999999999999999999999998086575293558792147769514 87
M = 5.05×10^{49} S = 100x75 D = 1000 AGDP ZEN

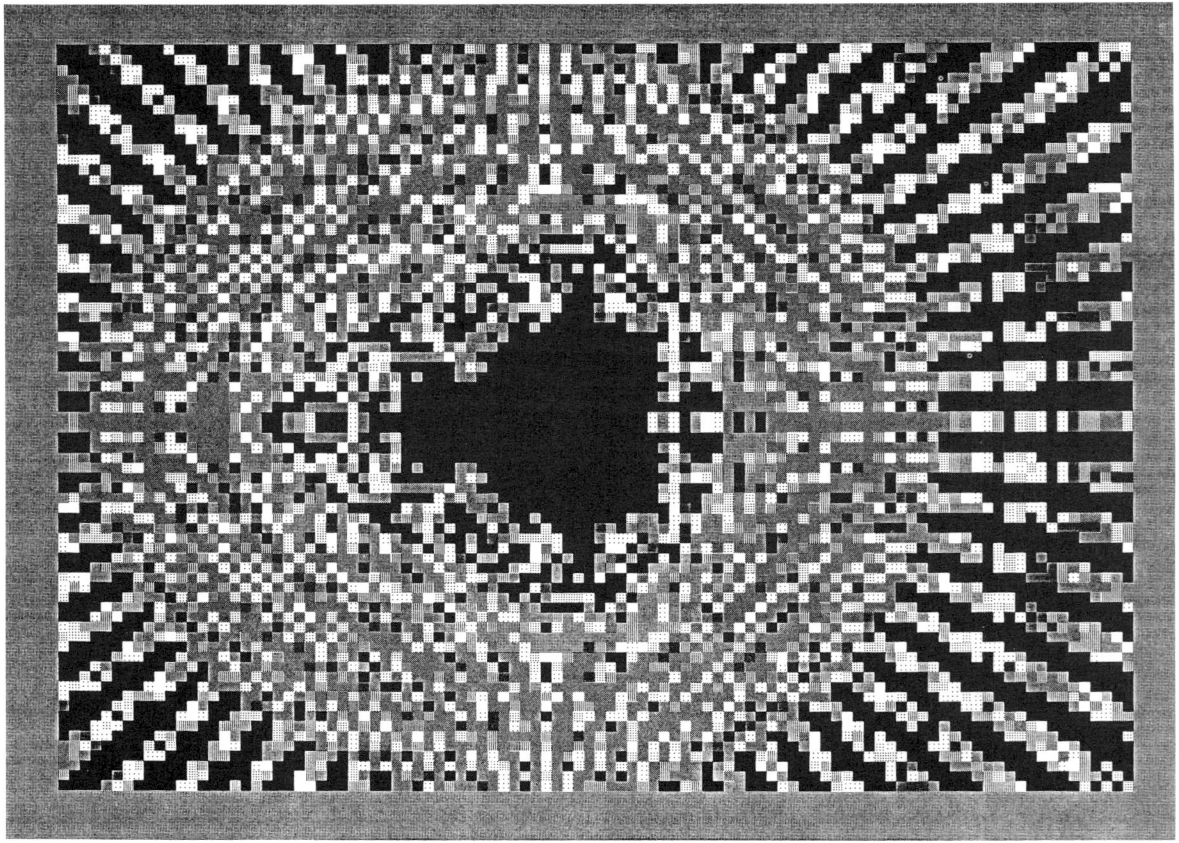

Cycle 44 R = -1.9999999999999999999999999995216438233896980369423791 87
M = $8.09 x 10^{50}$ S = 100x75 D = 1000 AR AT

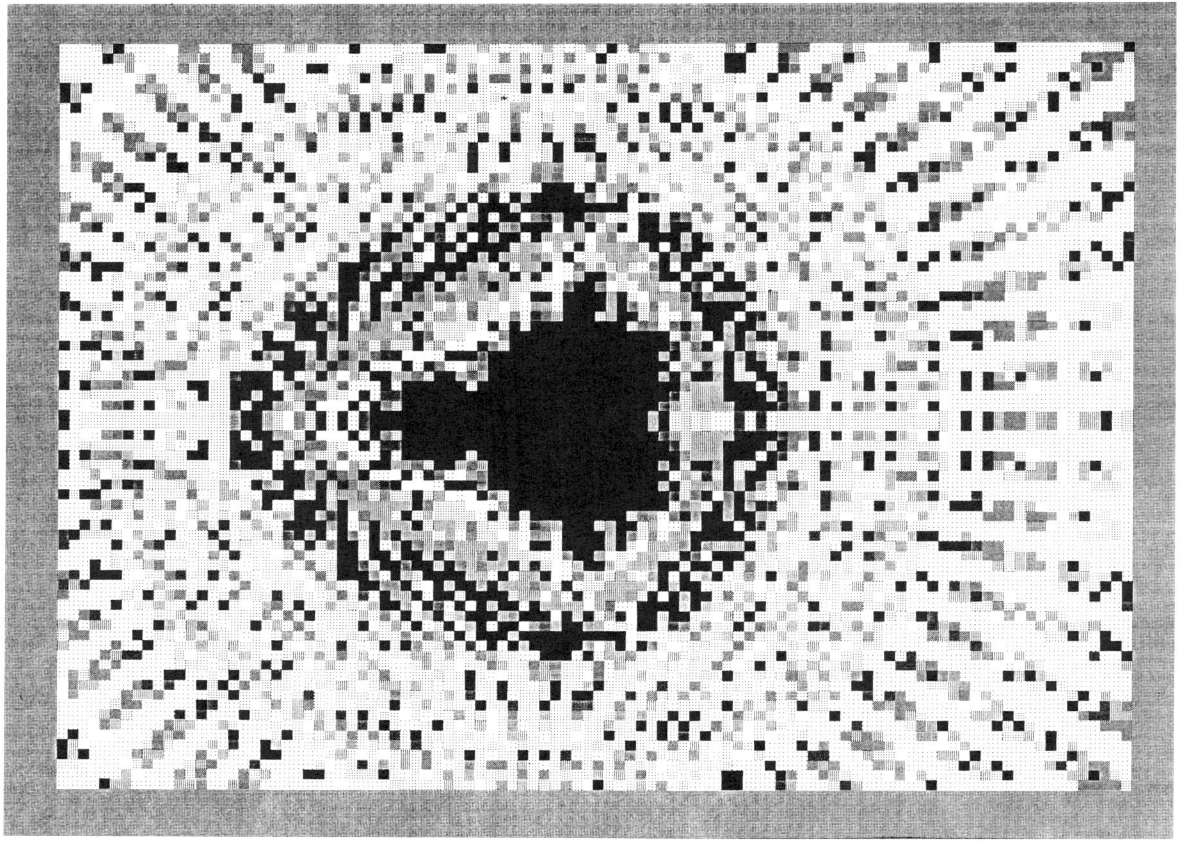

Cycle 45 R = -1.999999999999999999999999999880410955847424509235559482681
M = $1.29 x 10^{52}$ S = 100x75 D = 1000 AGDP ZEN

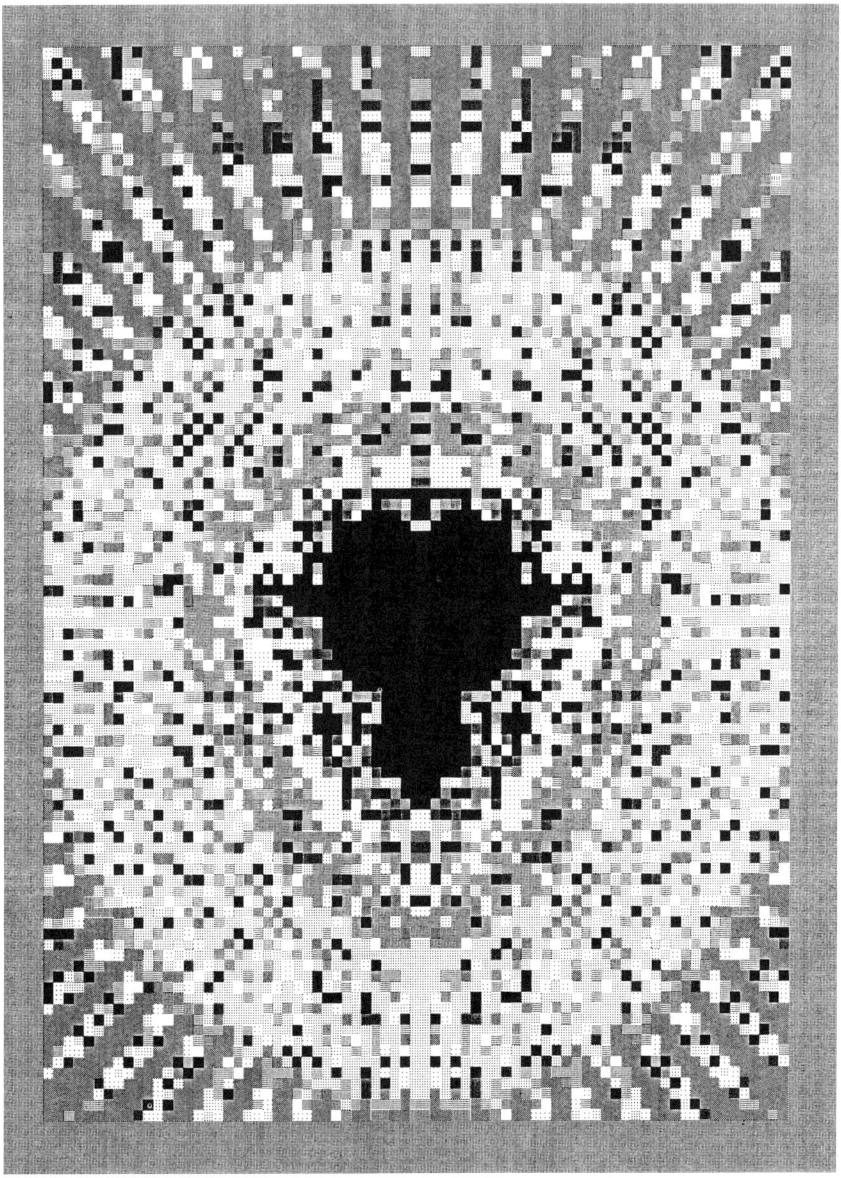

Cycle 46 R = -1.999999999999999999999970102738961856127308898708625
M = 2.07×10^{53} S = 100x75 D = 1000 AR AT

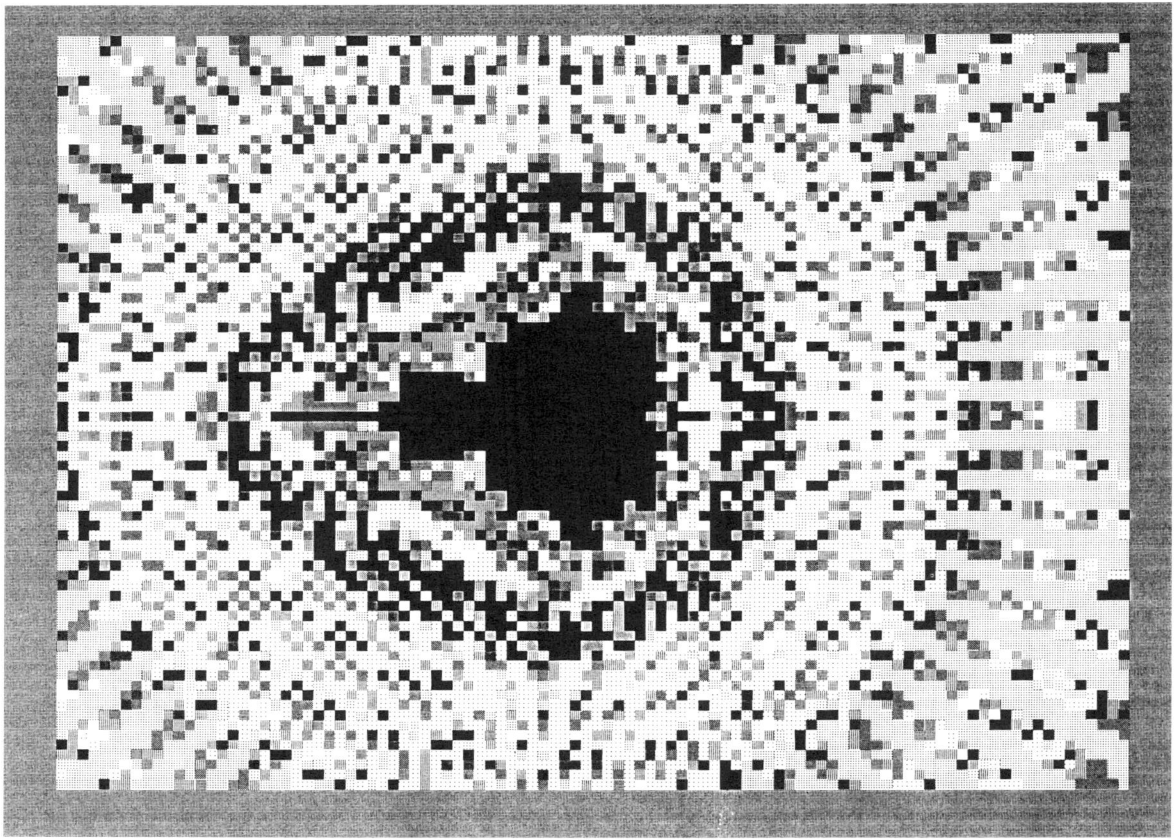

Cycle 47 R = -1.999999999999999999999999999999252568474046403182722246772793
M = $3.31 x 10^{54}$ S = 100x76 D = 1000 AGDP ZEN

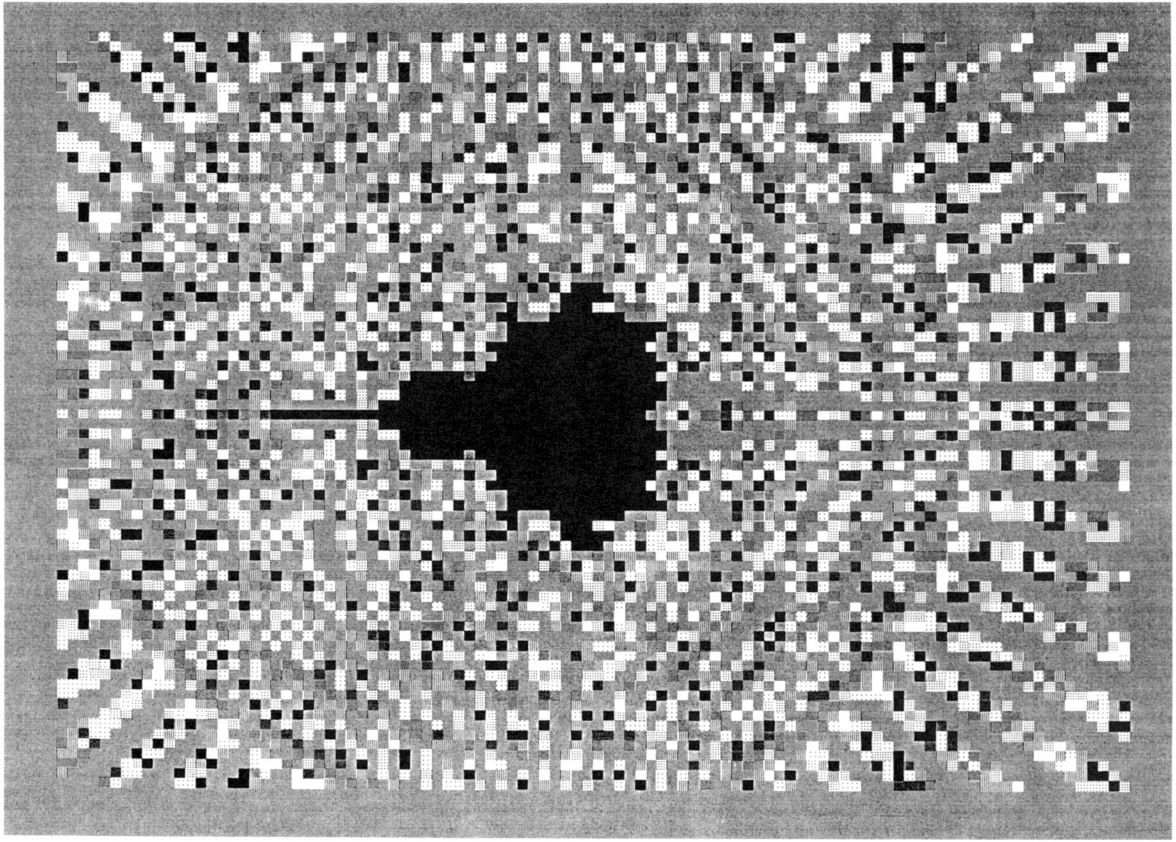

Cycle 48 R = -1.999999999999999999999999999999813142118511600795680616932770
M = 5.30×10^{55} S = 100x76 D = 1000 AGDP ZEN

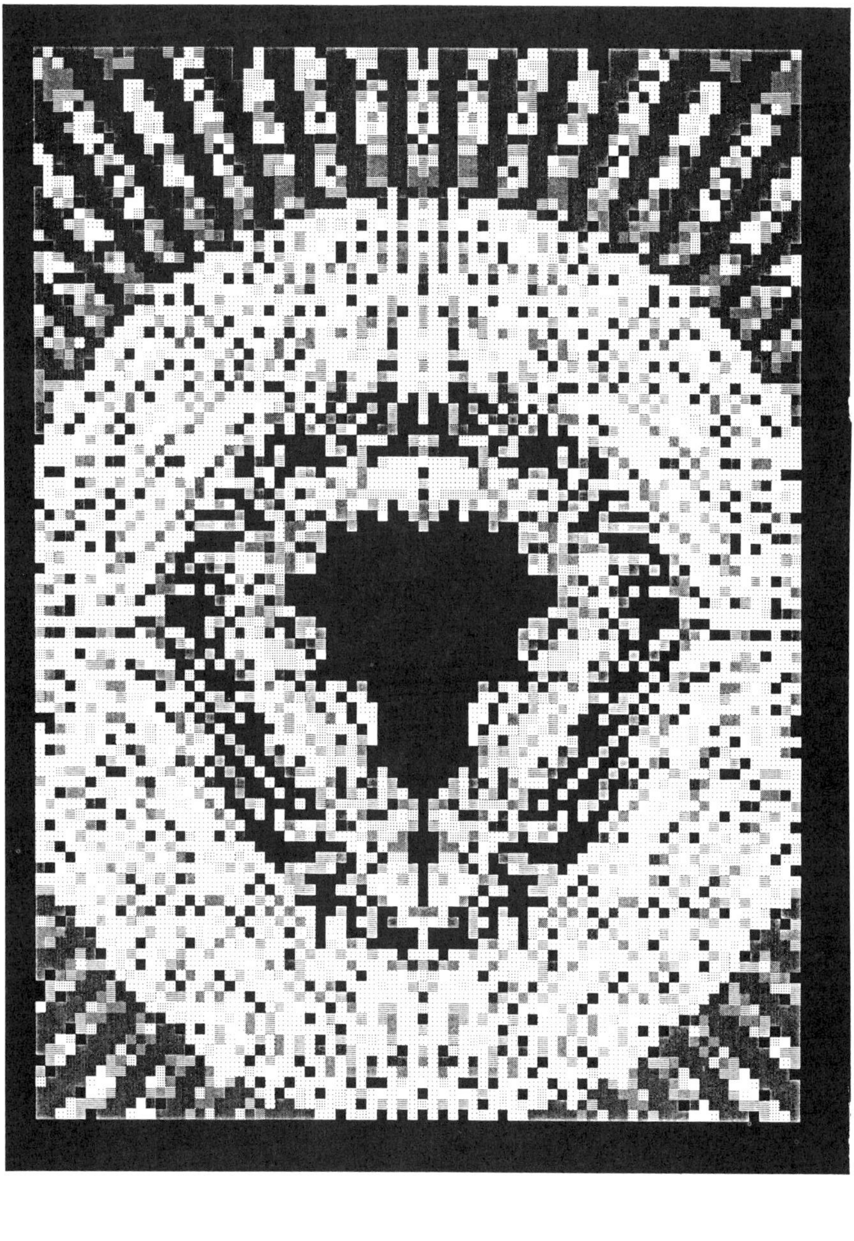

Cycle 49 R = -1.999999999999999999999999999995328552962790019892015423332428
M = $8.48x10^{56}$ S = 100x76 D = 1000 AGDP ZEN

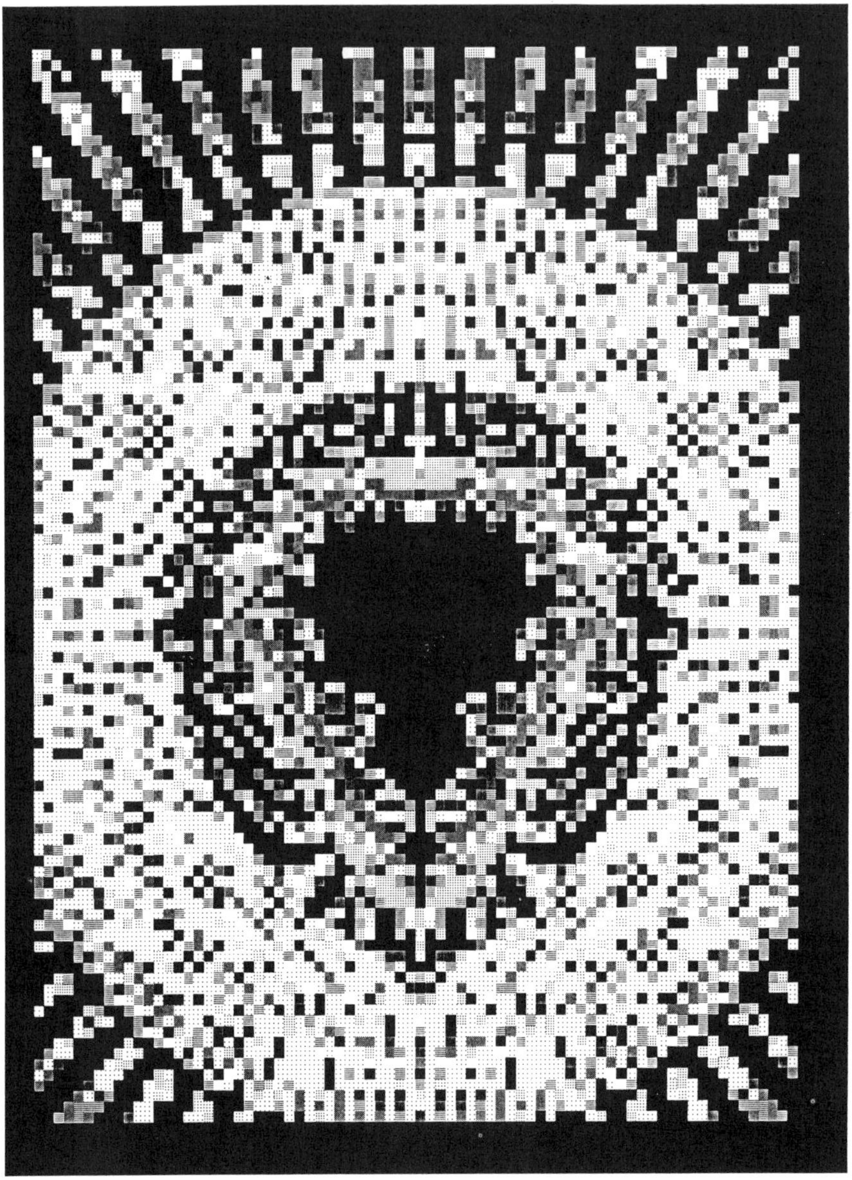

Cycle 50 R = -1.9999999999999999999999999988321382406975049730038558313920
M = 1.36×10^{58} S = 100x76 D = 1000 AGDP ZEN

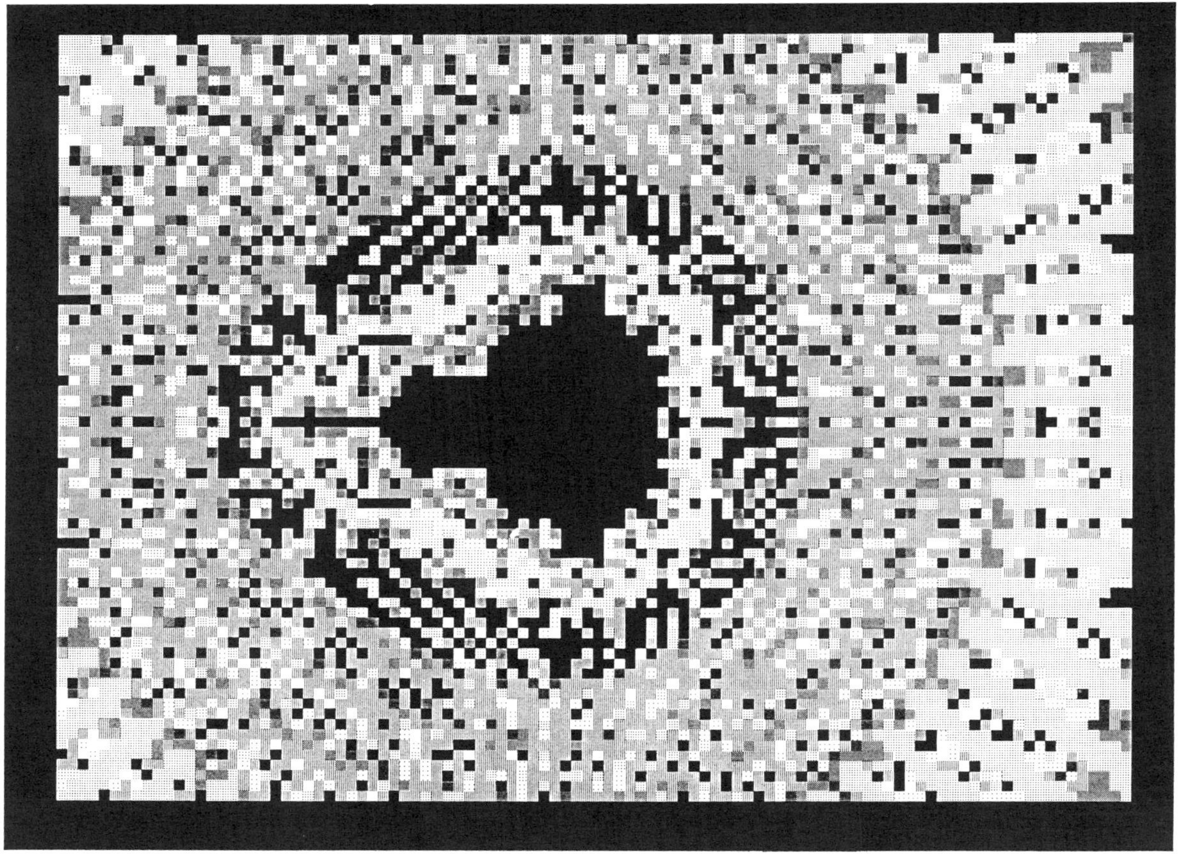

Cycle 51 R = -1.99999999999999999999999999999999708034560174376243250963957868 48
M = 2.17×10^{59} S = 100x76 D = 1000 AGDP ZEN

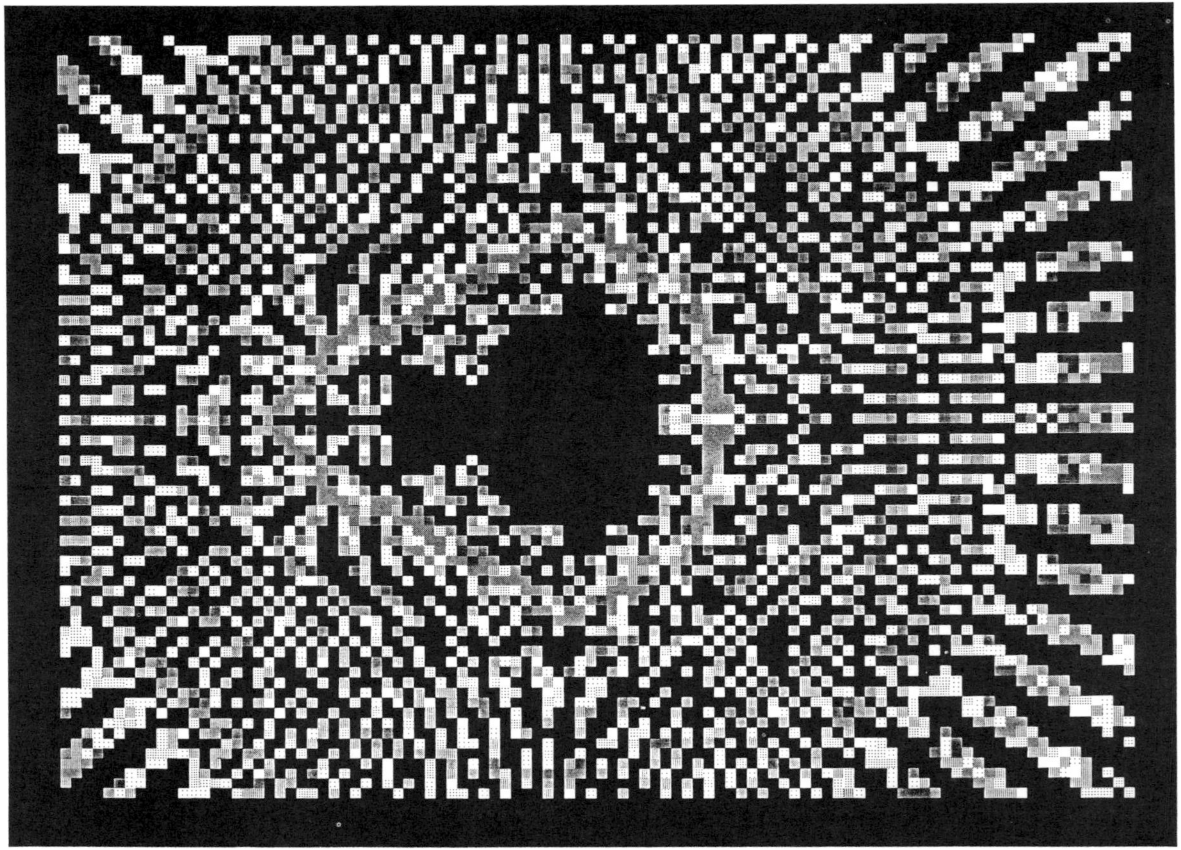

Cycle 52 R = -1.99999999999999999999999999999999927008640043594060812740989468427
M = $3.47x10^{60}$ S = 100x76 D = 1000 AGDP ZEN

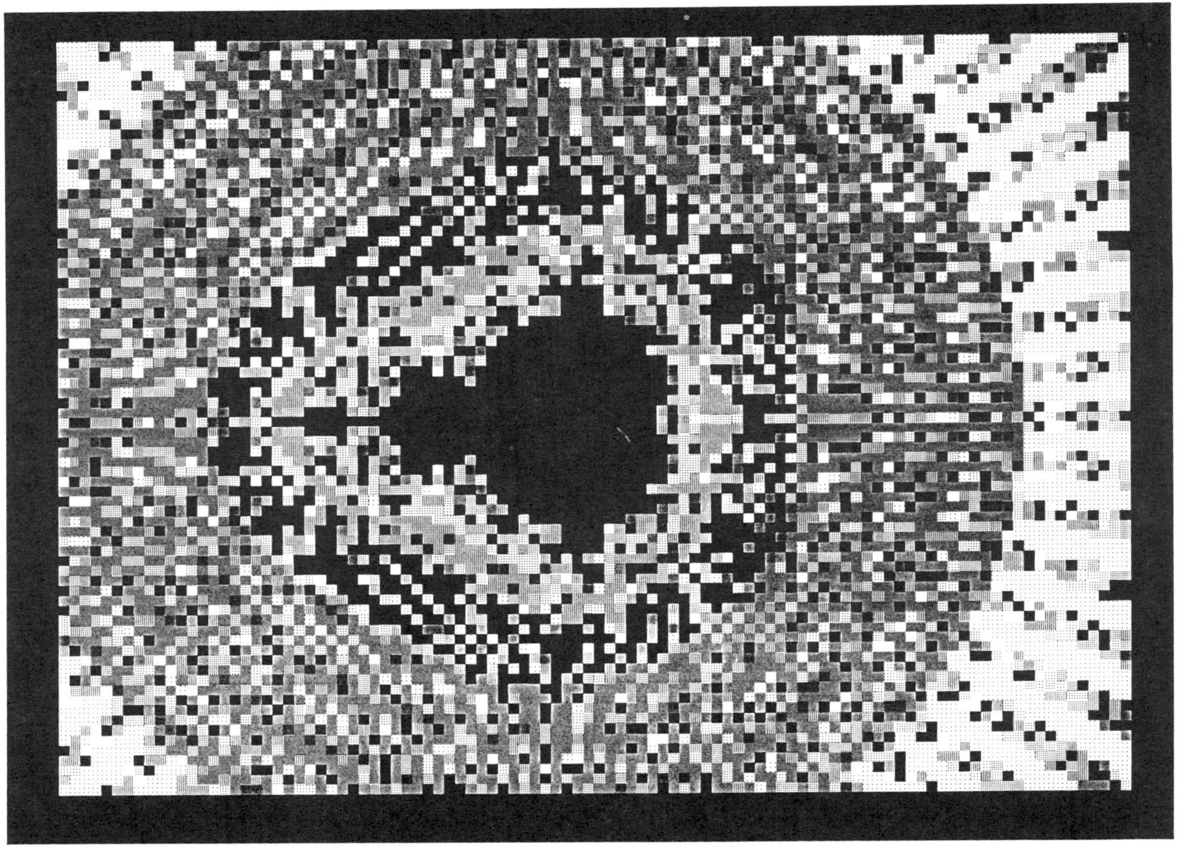

Cycle 53 R = -1.9999999999999999999999999999999999817521600108985152031852473671900
M = 5.56×10^{61} S = 100x 76 D = 1000 AGDP ZEN

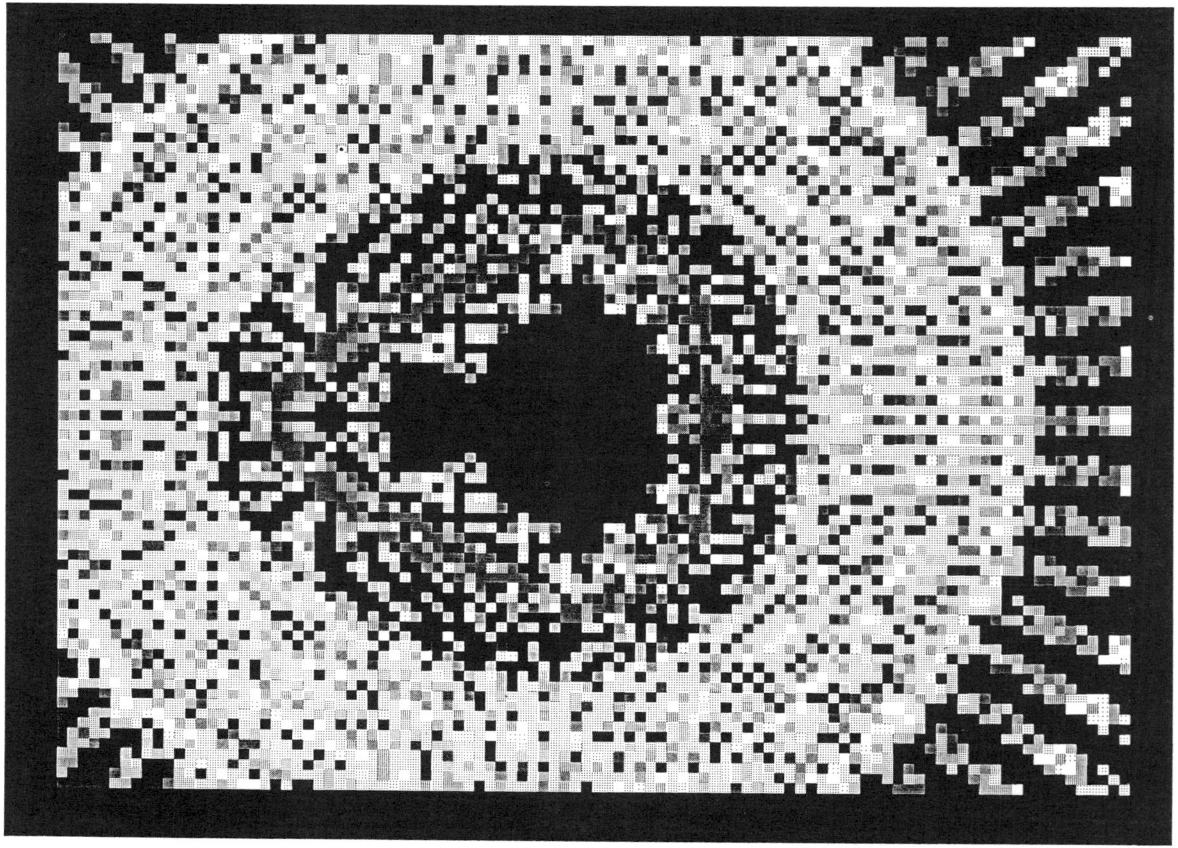

Cycle 54 R = -1.99999999999999999999999999999999999543804000272462880079631184180281 M = 7.11×10^{62} S = 100x76 D = 1000 AGDP ZEN

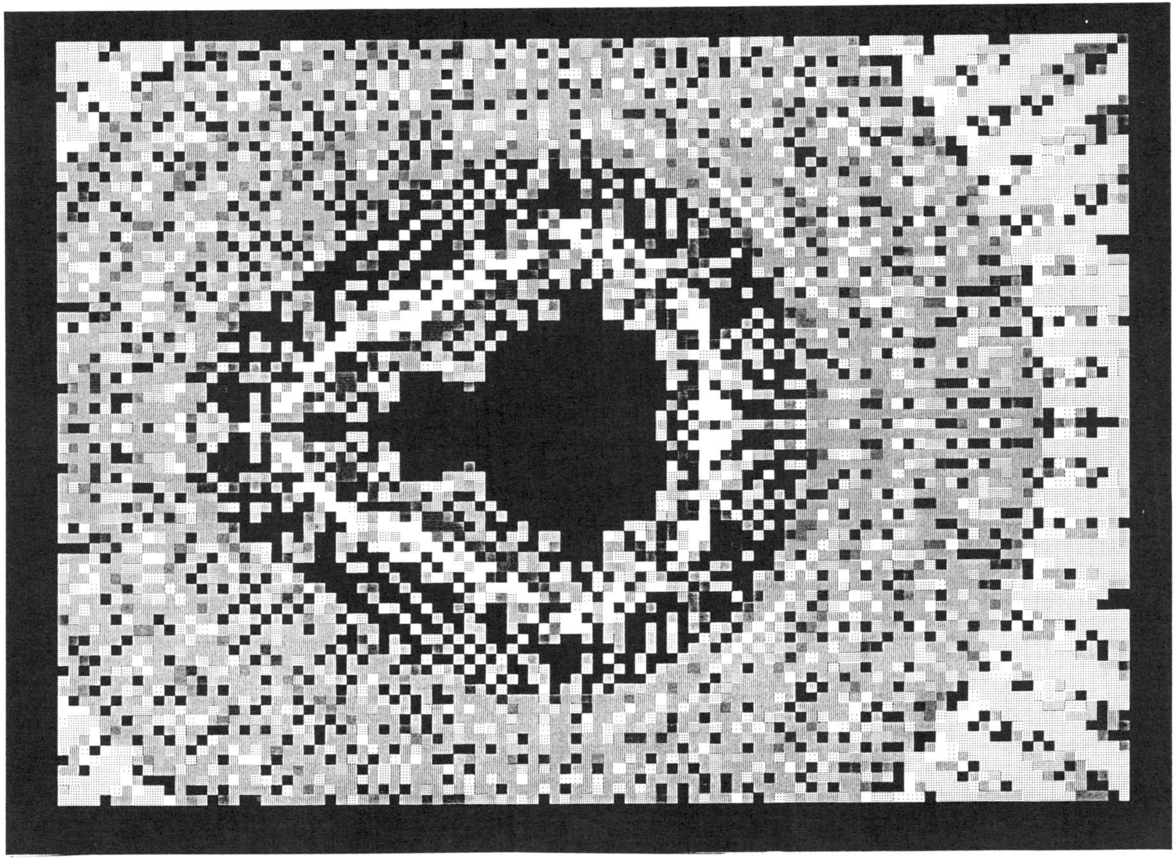

Cycle 55 R = -1.99999999999999999999999999999988595100006811572001990779604510
41 M = 1.14×10^{64} S = 100x76 D = 1000 AGDP ZEN

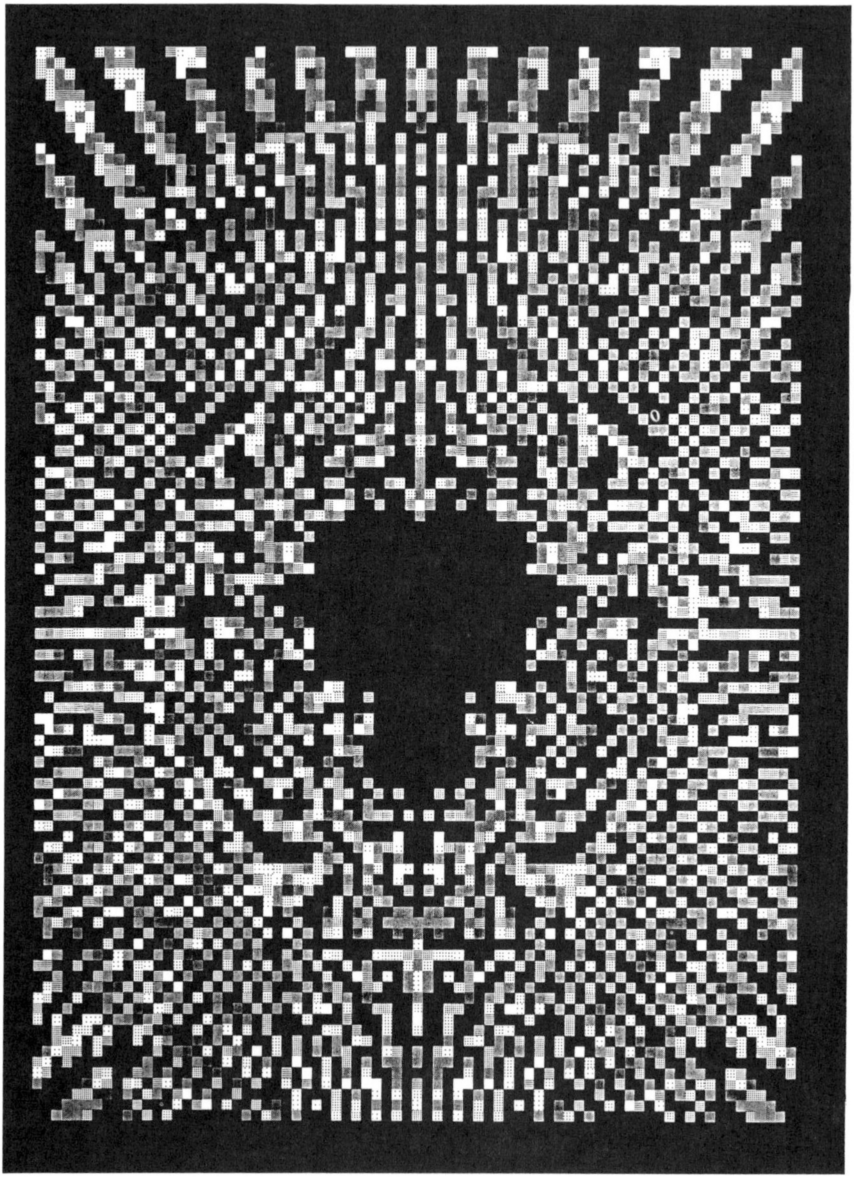

Cycle 56 R = -1.99999999999999999999999999999999714877500170289300049769490112 7819 M = $1.82x10^{65}$ S = 100x76 D = 1000 AGDP ZEN

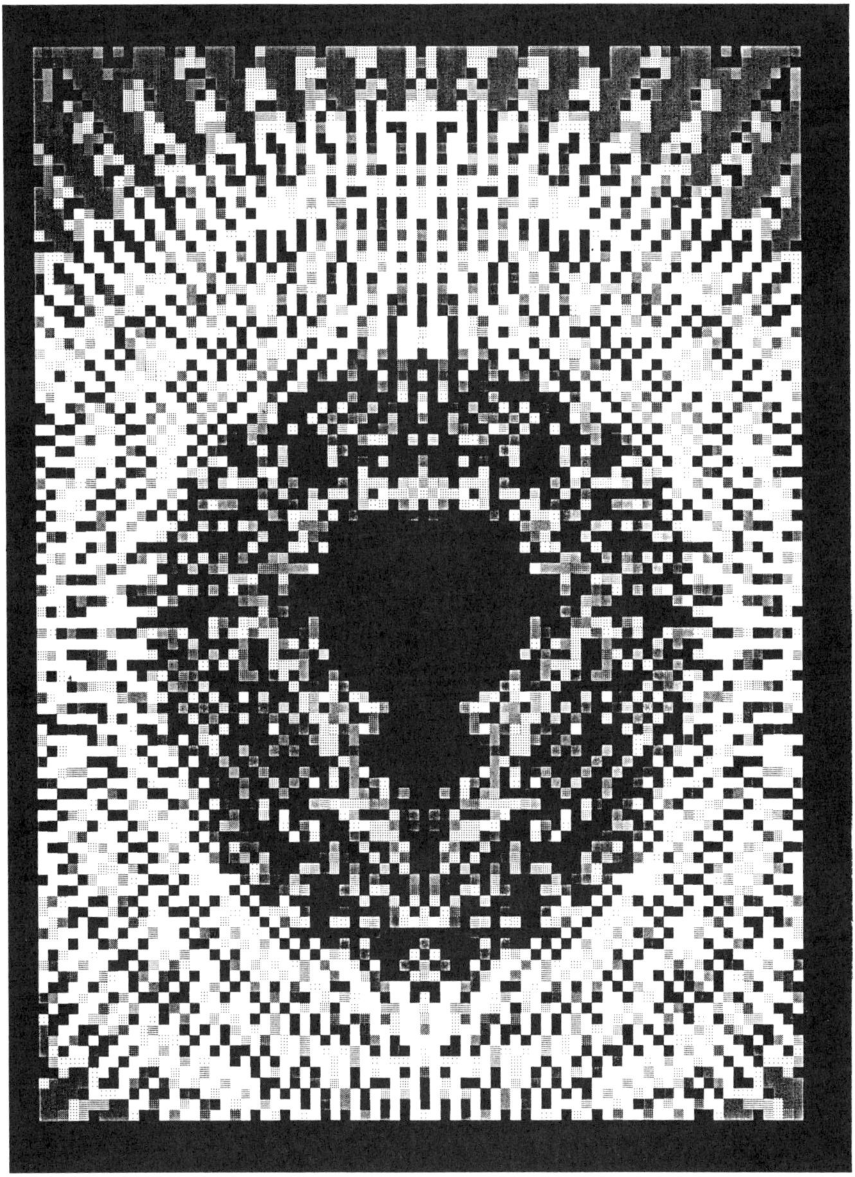

Cycle 57 R = -1.999999999999999999999999999999999287193750425723225012442372528 1
9684 M = 2.91×10^{66} S = 100x76 D = 1000 AGDP ZEN

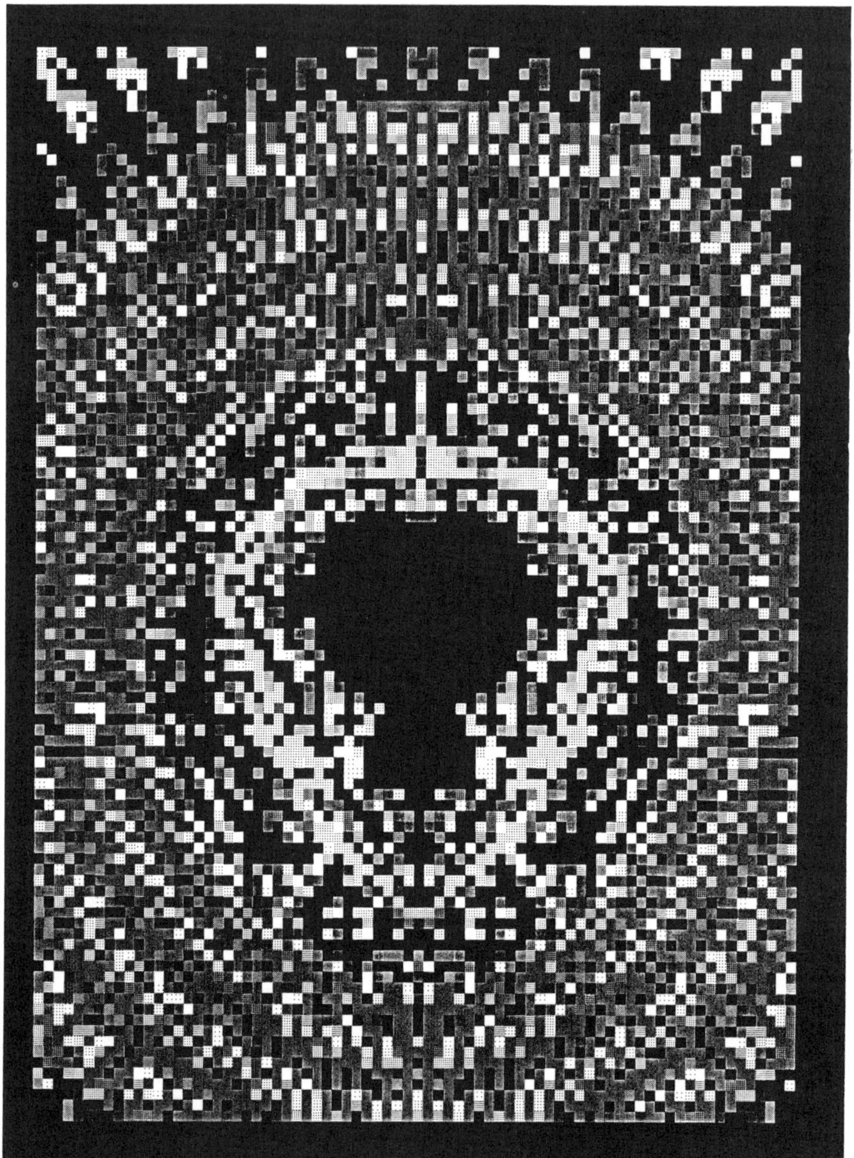

Cycle 58 R = -1.9999999999999999999999999999999982179843760643081253110593132049299 M = 4.66×10^{67} S = 100x76 D = 1000 AGDP ZEN

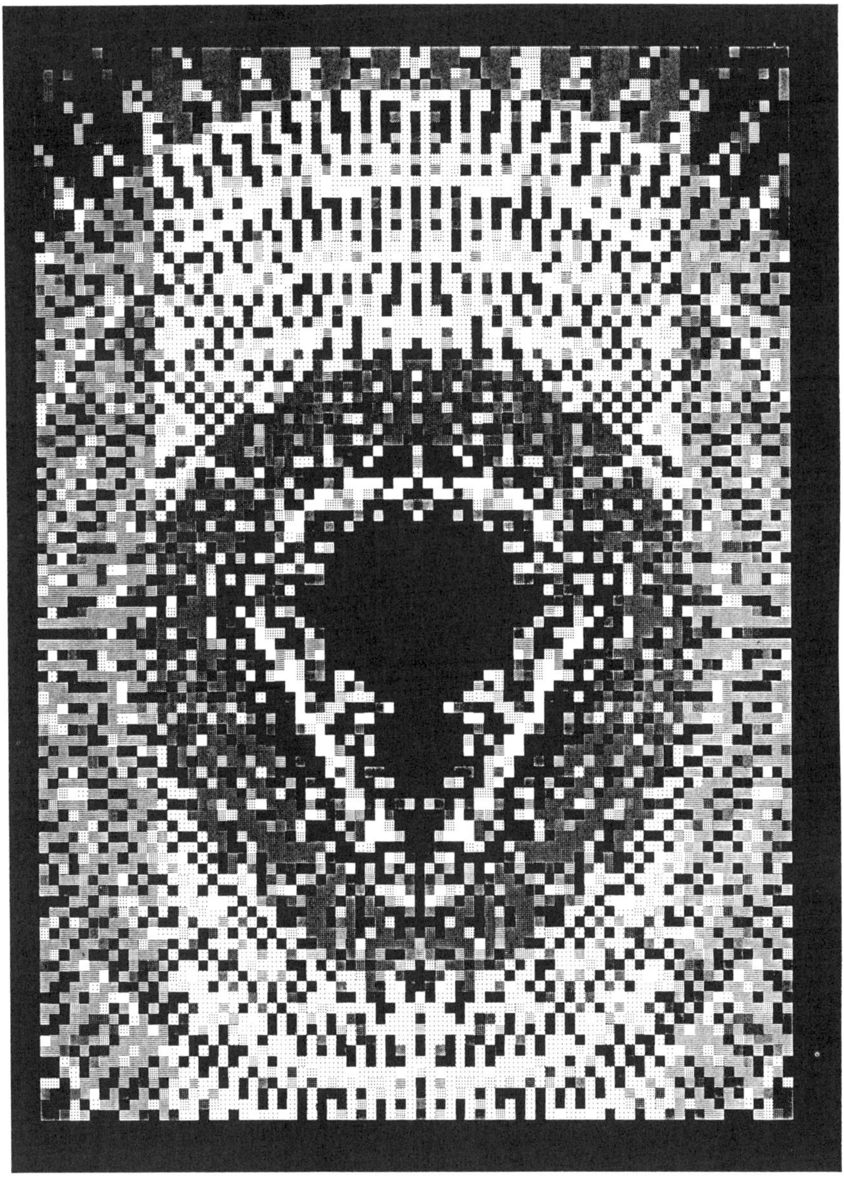

Cycle 59 R = -1.999999999999999999999999999999999955449609401607703132776482830
123304 M = 7.46×10^{68} S = 100x76 D = 1500 AGDP ZEN

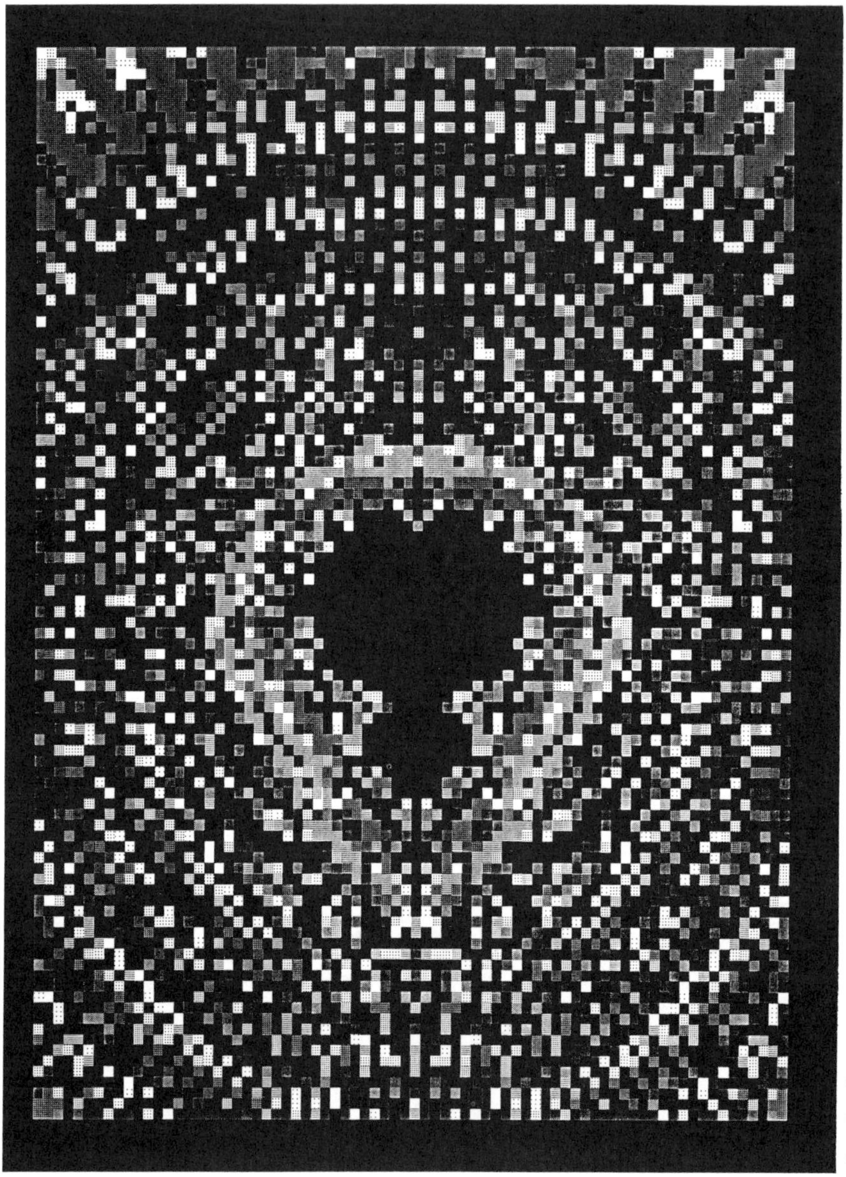

Cycle 60 R = -1.999999999999999999999999999999888624023504019257831941207075308 2974 M = 1.19x10^70 S = 100x76 D = 1500 AGDP ZEN

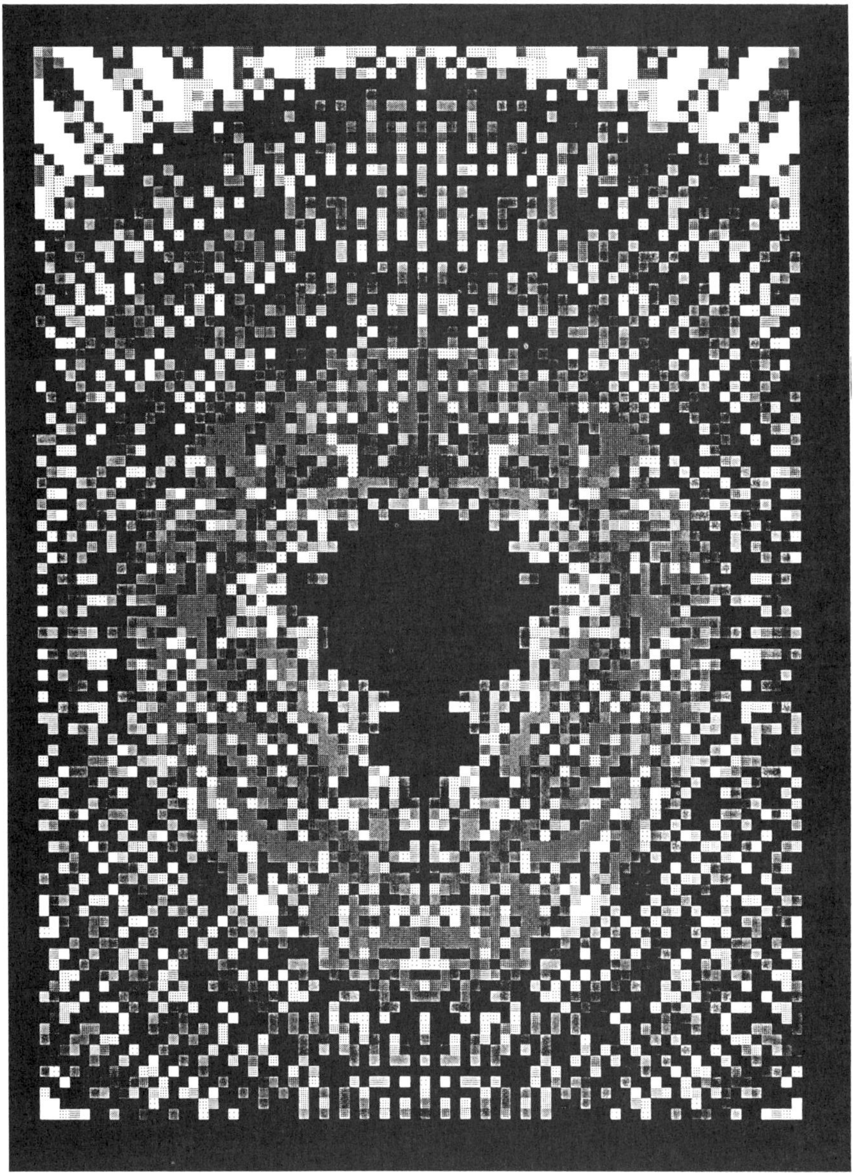

Cycle 61 R = -1.999999999999999999999999999999999999972156000587600481445798530176 882707661 M = 1.91x10^{71} S = 100x76 D = 1500 AGDP ZEN

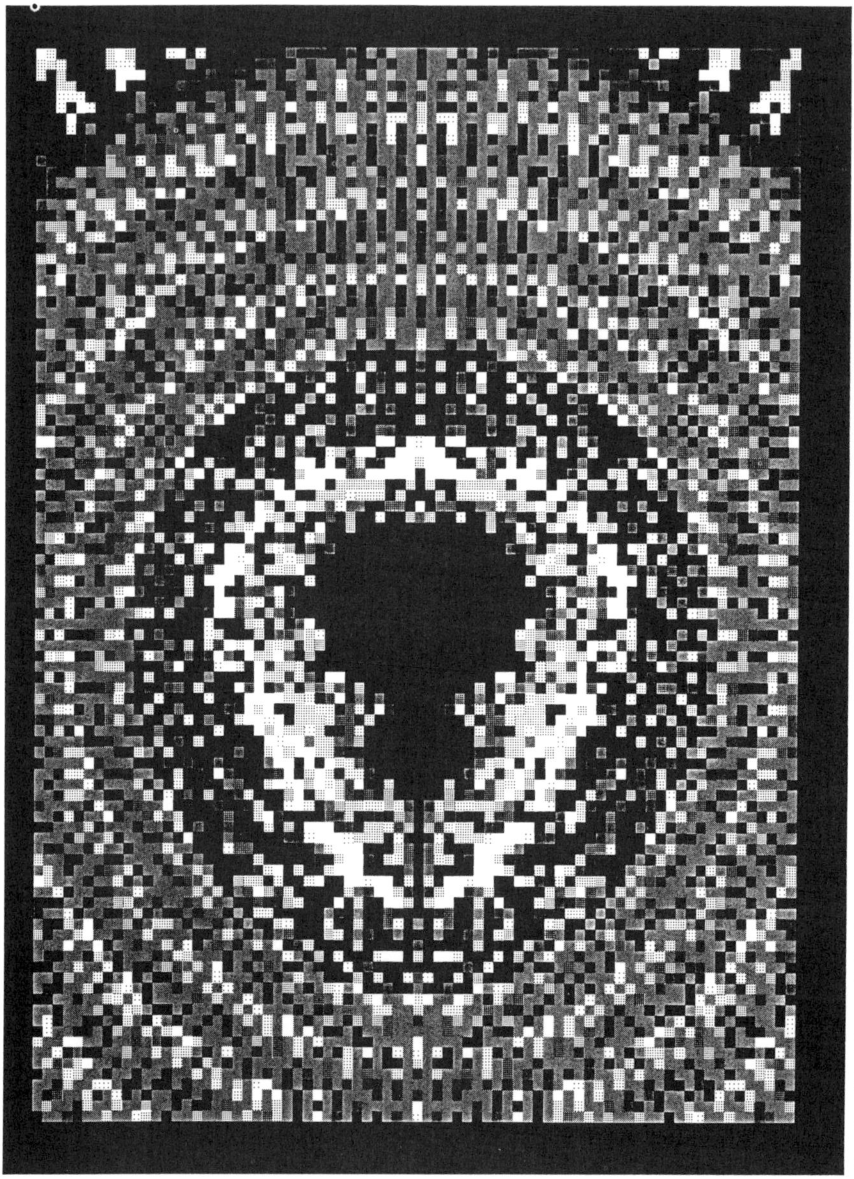

Cycle 62 R = -1.999999999999999999999999999999999999303900146900120361449632544 2206769296 M = $3.06x10^{72}$ S = 100x76 D = 1500 AGDP ZEN

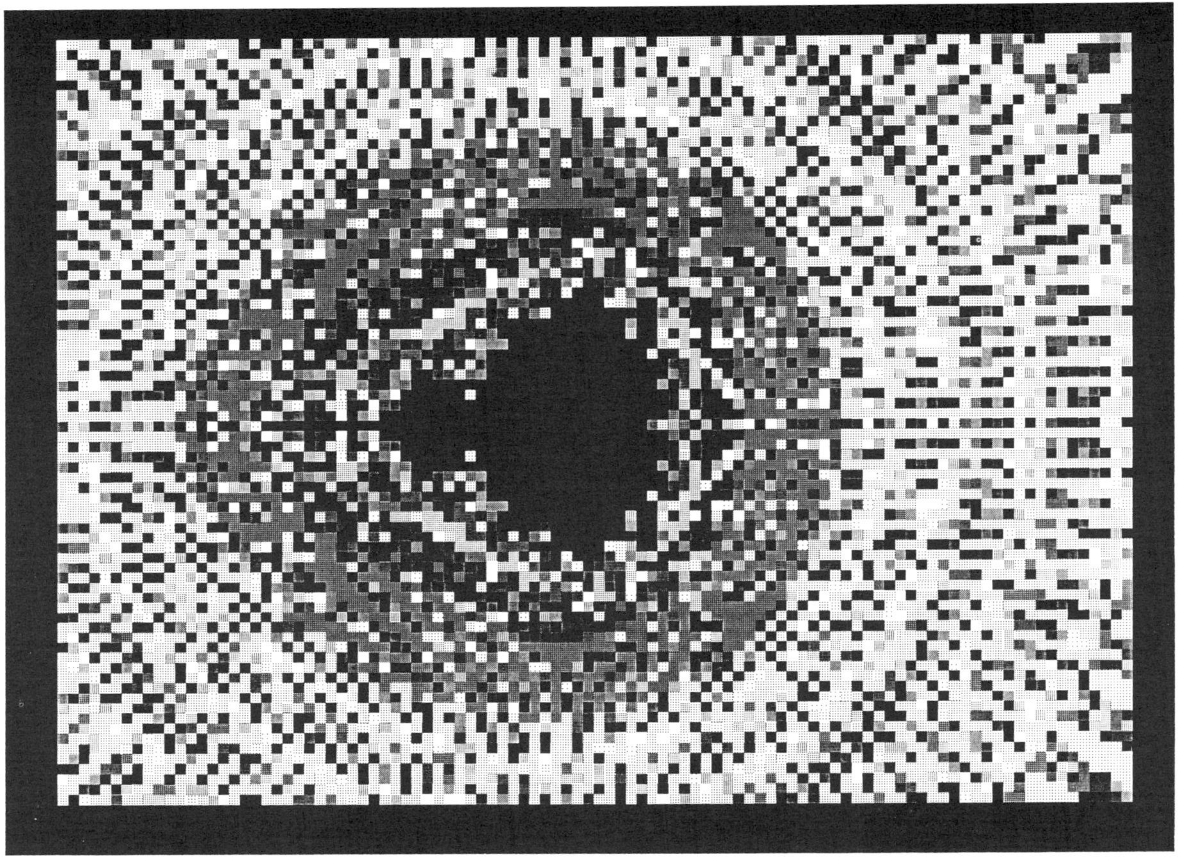

Cycle 63 R = -1.9999999999999999999999999999999999999825975036725030090362408136 05516923331 M = 4.88×10^{73} S = 100x76 D = 1500 AGDP ZEN UC

Cycle 64 R = -1.99999999999999999999999999999999999995649375918125752259060203401379230838 6 M = 9.78×10^{74} S = 100x76 D = 1500 AGDP ZEN UC

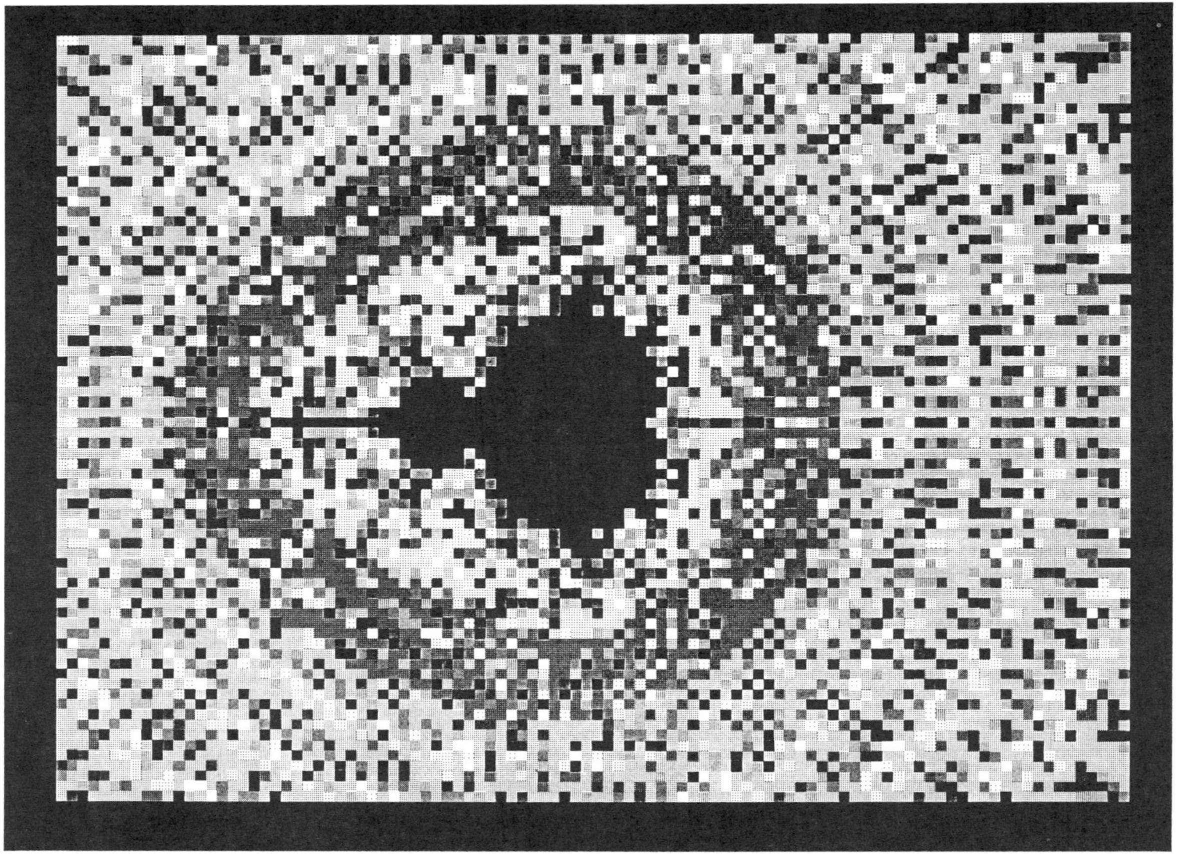

Cycle 65 R = -1.999999999999999999999999999999999999999999999891234397953143806476505085034480771002 2 M = $1.56x10^{76}$ S = 100x76 D = 1500 AGD ZEN UC

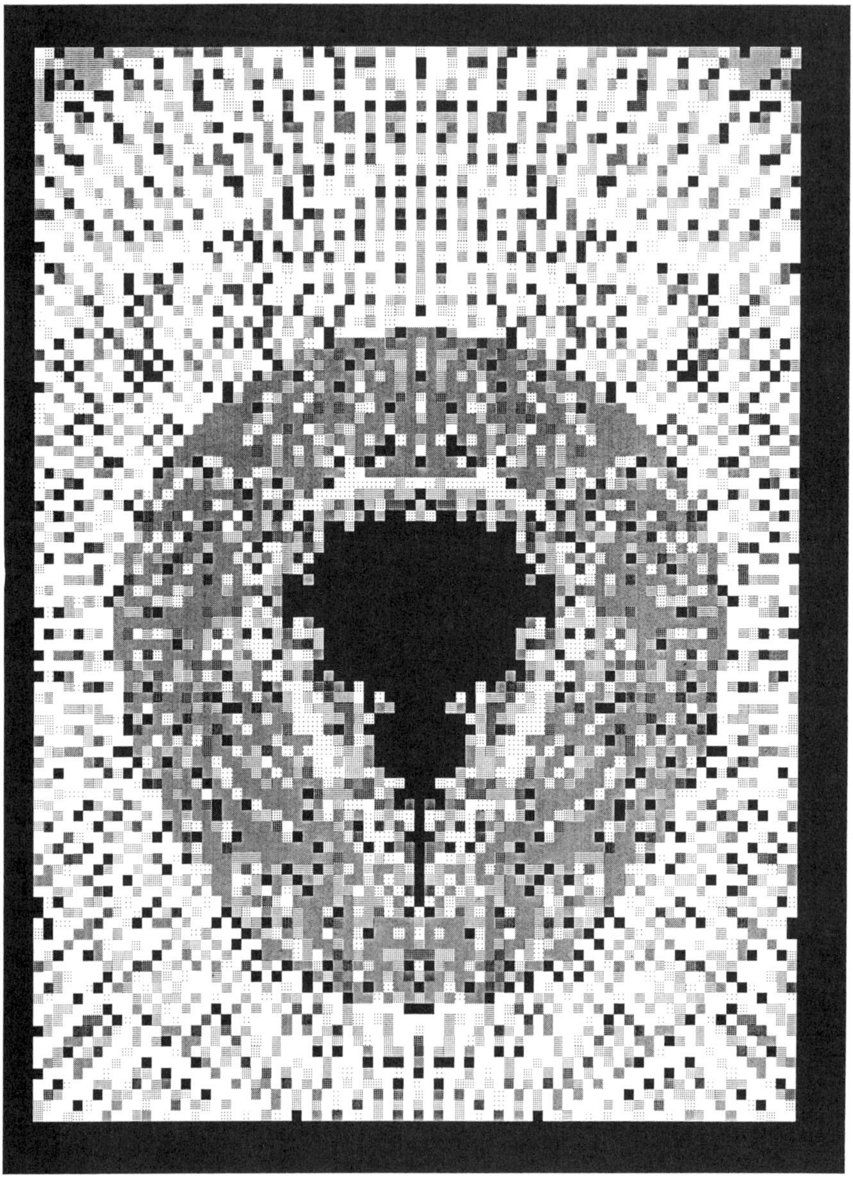

Cycle 66 R = -1.99999999999999999999999999999999999997280859948828595161912627125862019275289 M = 2.50x10^{77} S = 100x76 D = 1500 AGDP ZEN UC

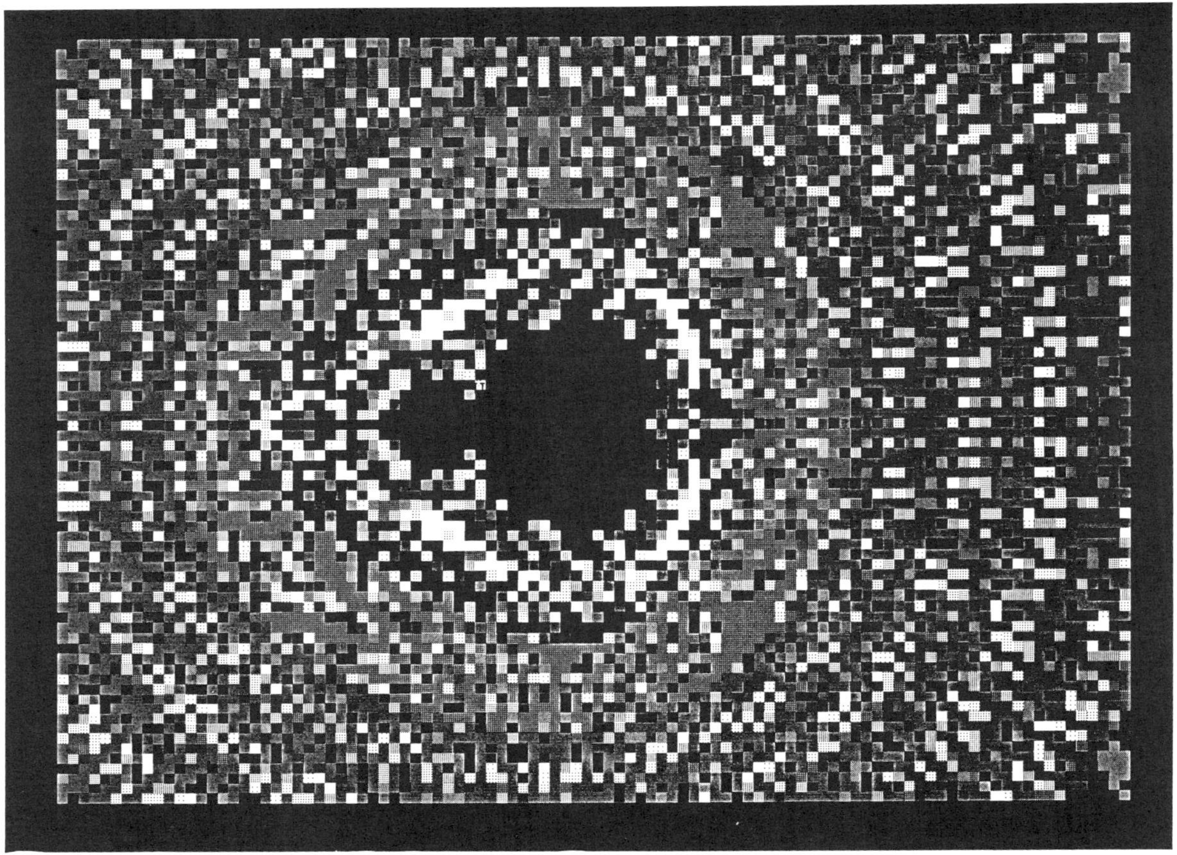

Cycle 67 R = -1.999999999999999999999999999999999999999320214987207148790478156
78146550481883720 M = 4.00×10^{78} S = 100x76 D = 1500 AGDP ZEN UC

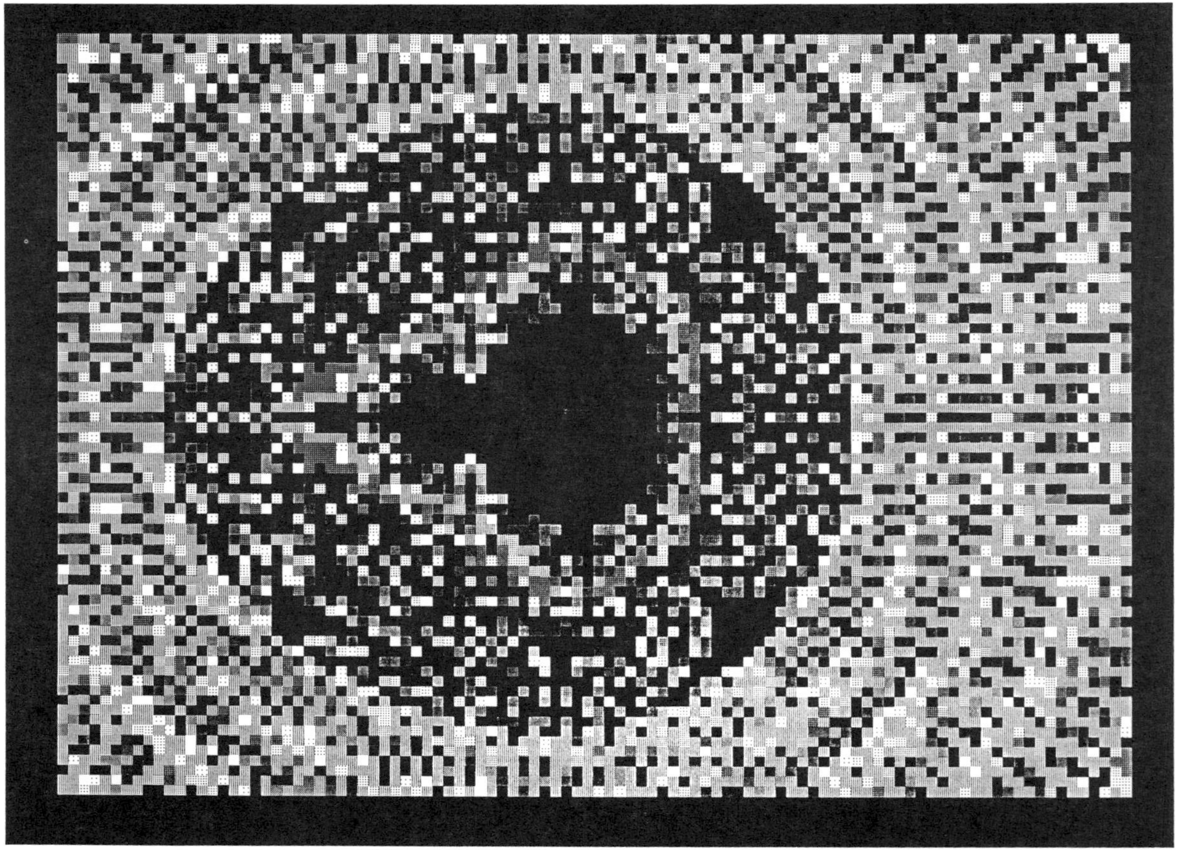

Cycle 68 R = -1.999999999999999999999999999999999999999830053746801787197619539 195366376204710240 M = 6.41x10^{79} S = 100x76 D = 1500 AGDP ZEN UC

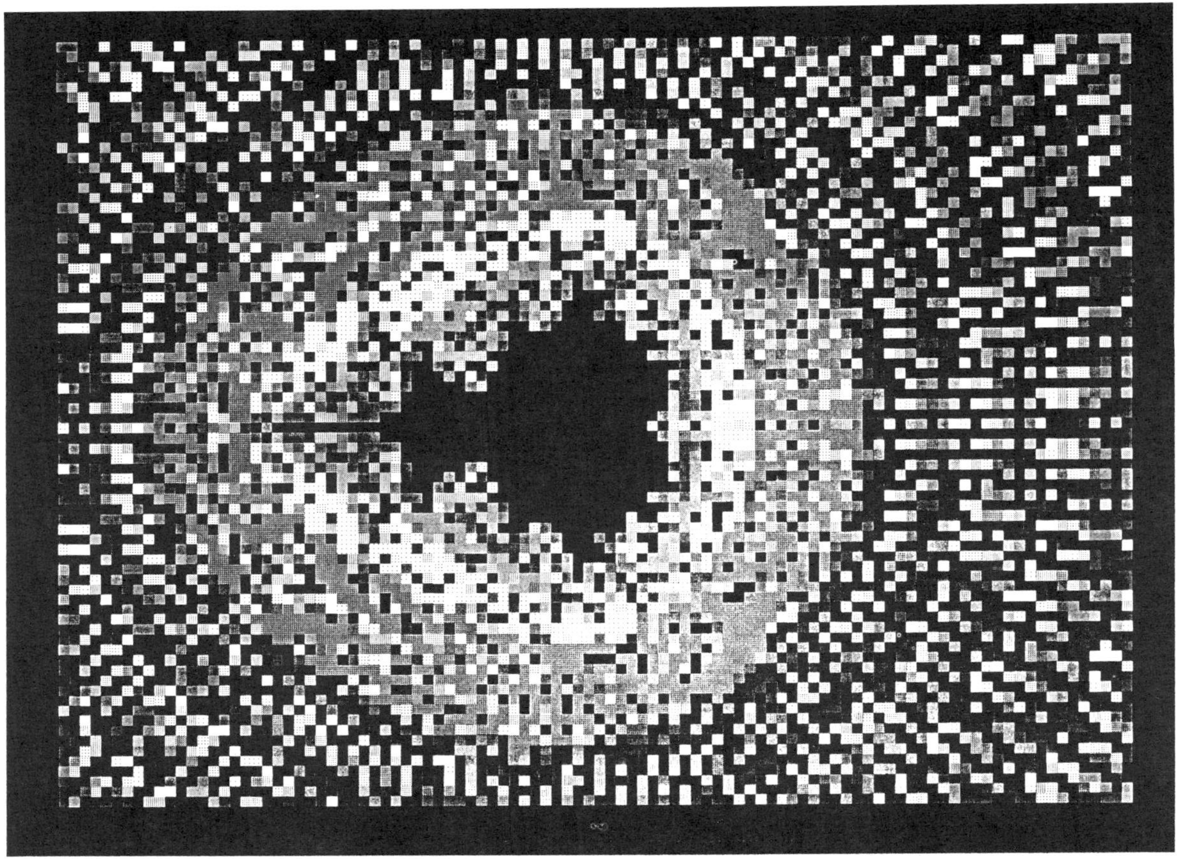

Cycle 69 R = -1.99999999999999999999999999999999999999957513436700446799404884 798841594051117776196 M = 1.03×10^{80} S = 100x76 D = 1500 AGDP ZEN UC

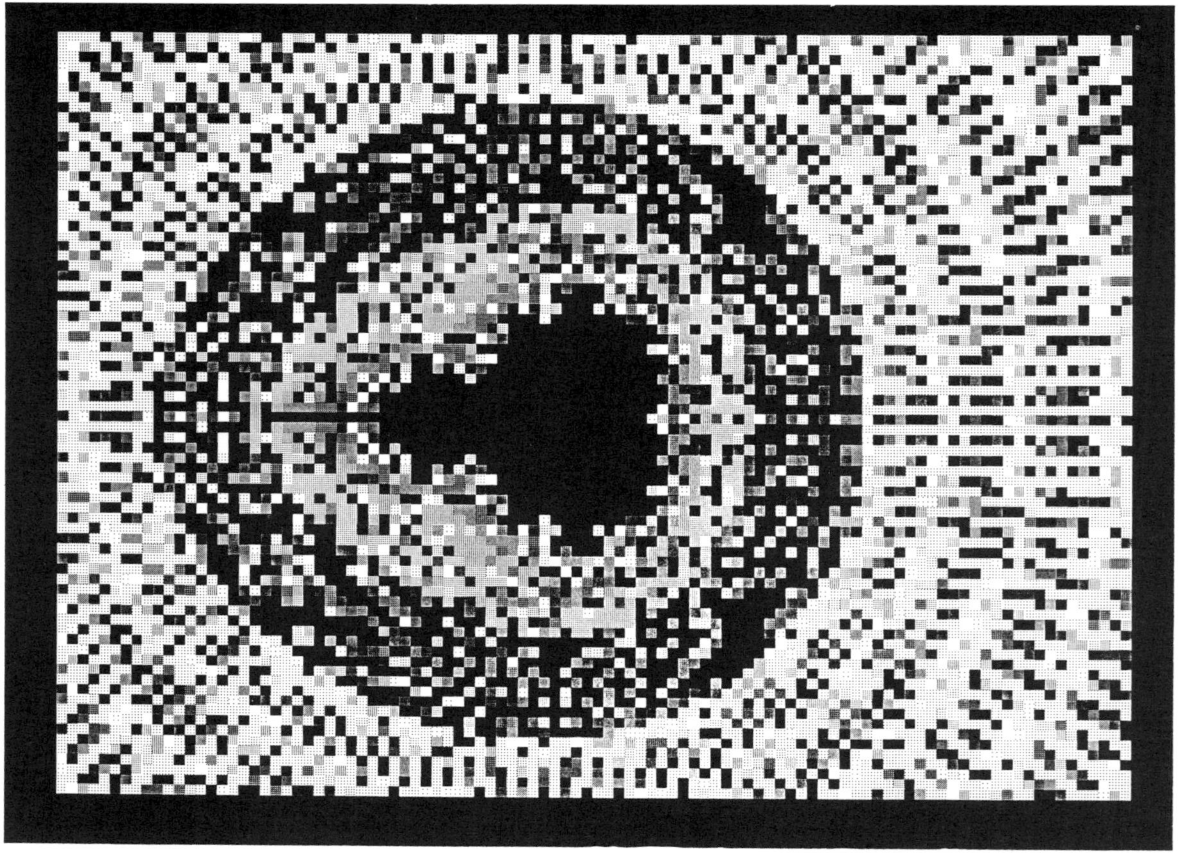

Cycle 70 R = -1.999999999999999999999999999999999999999999893783591751116998512211997103985127944 08699 M = $1.64x10^{82}$ S = 100x76 D = 1500 AGDP ZEN UC

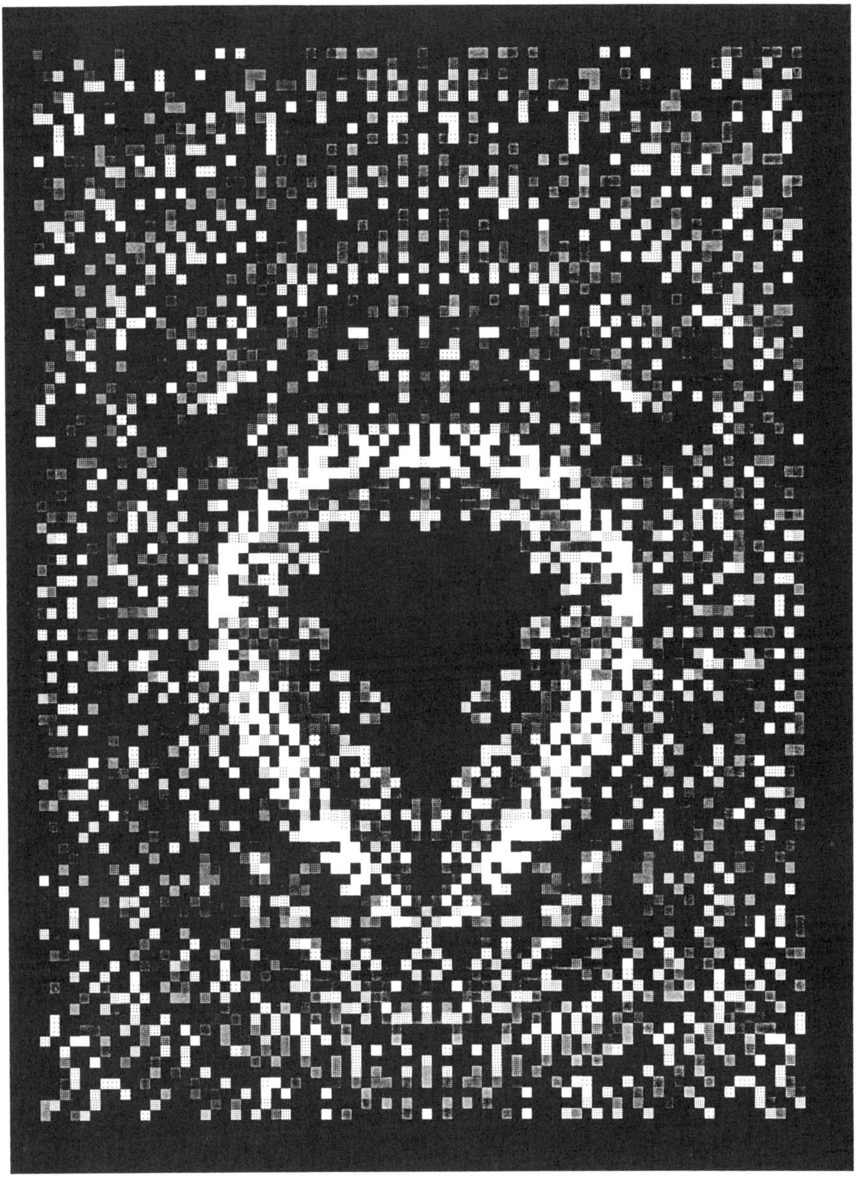

Cycle 71 R = -1.999999999999999999999999999999999999999999999999997344589793777924962805299927599628198 6024147 M = 2.62×10^{83} S = 100x76 D = 1500 AR VAX

Cycle 72 R = -1.99999999999999999999999999999999999999993361474484444812407013249818999070496 5061889 M = 4.20×10^{84} S = 100x76 D = 1500 AR VAX

Cycle 73 R = -1.999999999999999999999999999999999999999834036862111120310175331245474976762412655687 M = 6.72×10^{85} S = 100x76 D = 1500 AR VAX

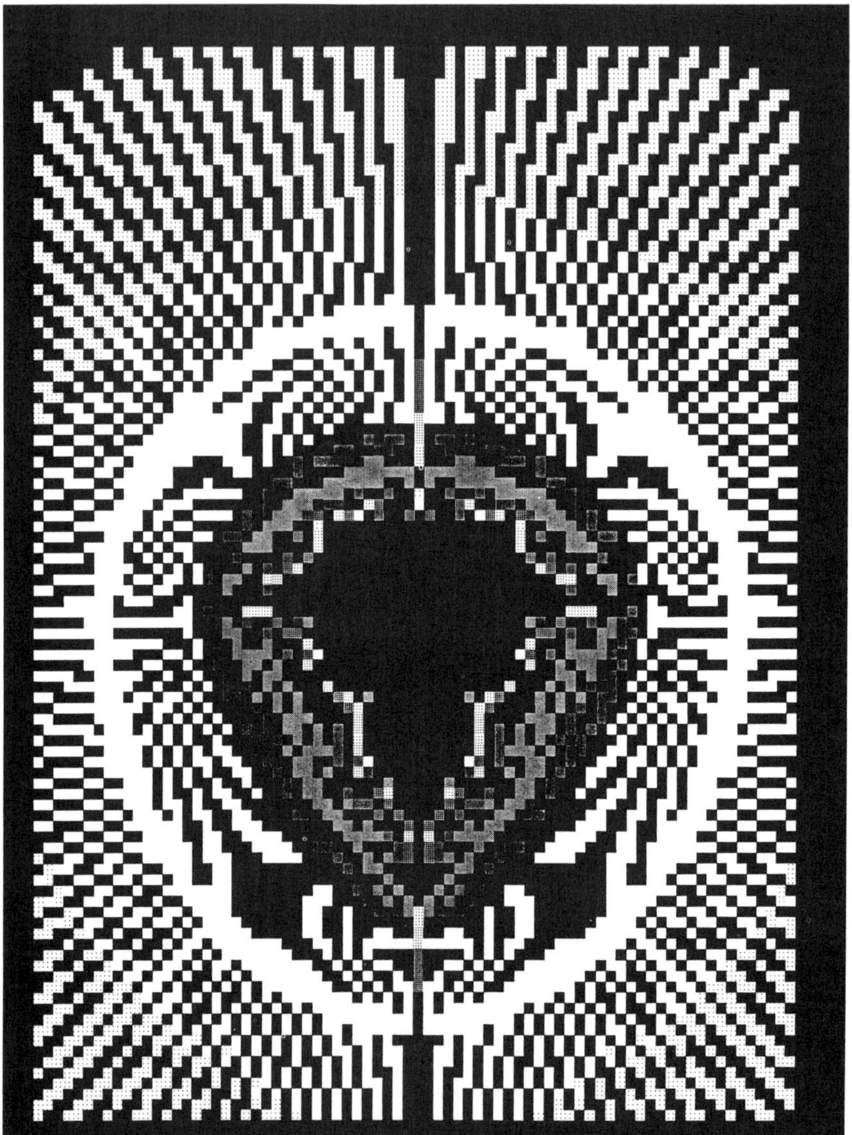

Cycle 74 R = -1.999999999999999999999999999999999999999999585092155277800775438328113687441906 031639829 M = 1.08x10^{87} S = 100x76 D = 1500 AR VAX

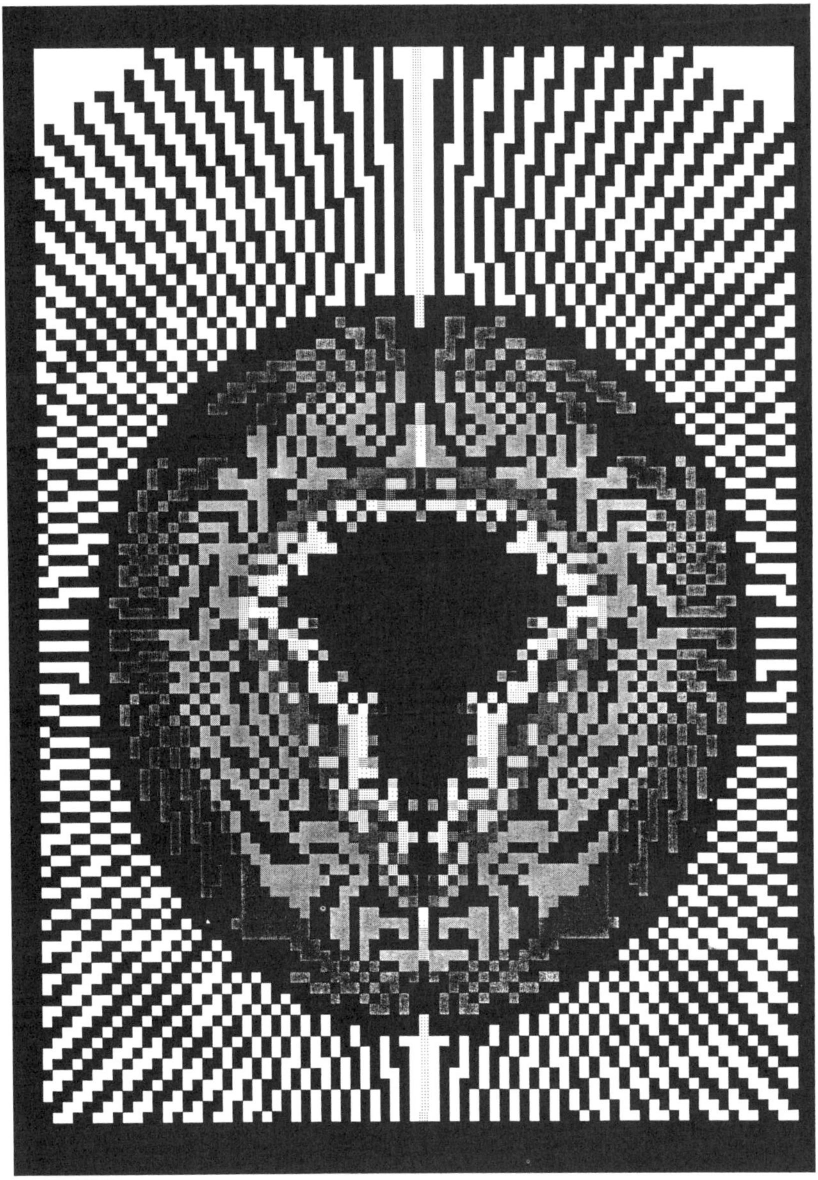

Cycle 75 R = -1.9999999999999999999999999999999999999999999896273038819450193859582028421860476 5079099961 M = 1.72×10^{88} S = 100x76 D = 1500 AR VAX

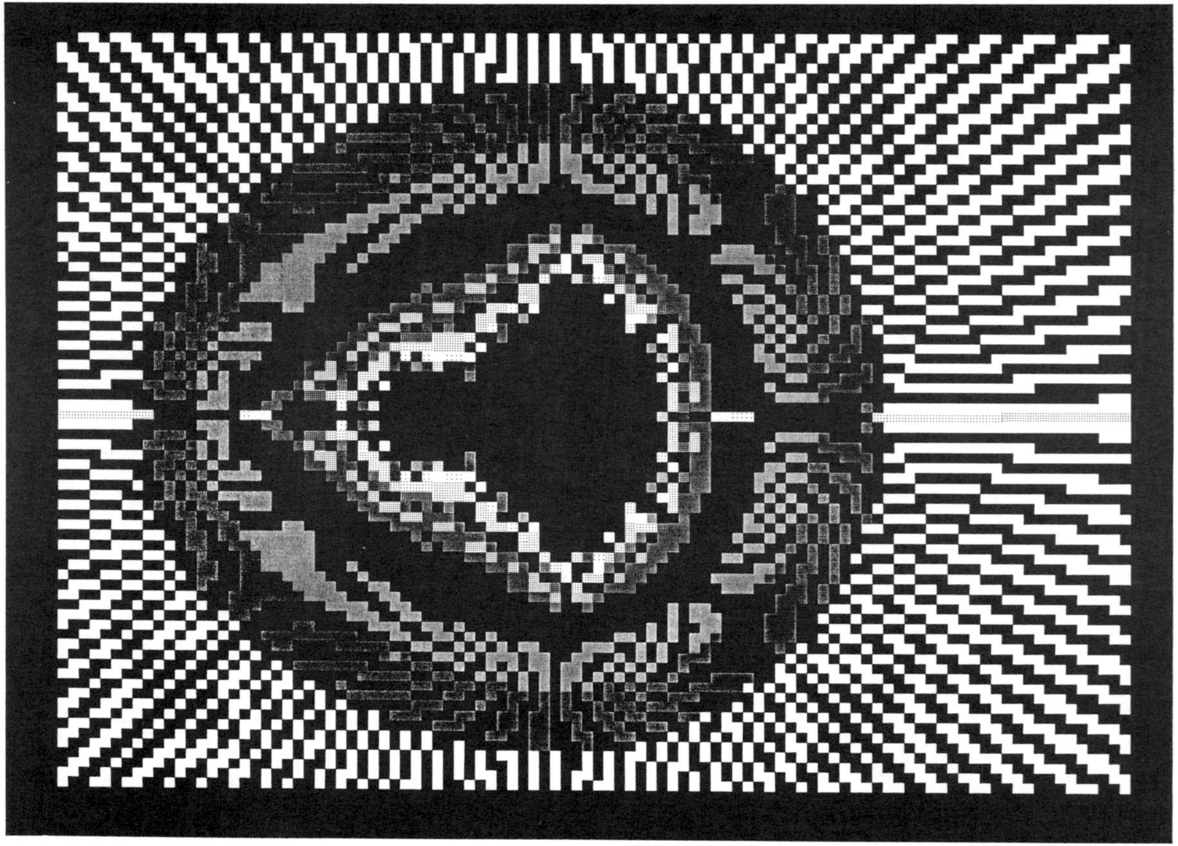

Cycle 76 R = -1.99999999999999999999999999999999999999999740682597048625484648955071054651191269775014 9 M = 2.75×10^{89} S = 100x76 D = 1500 AR VAX

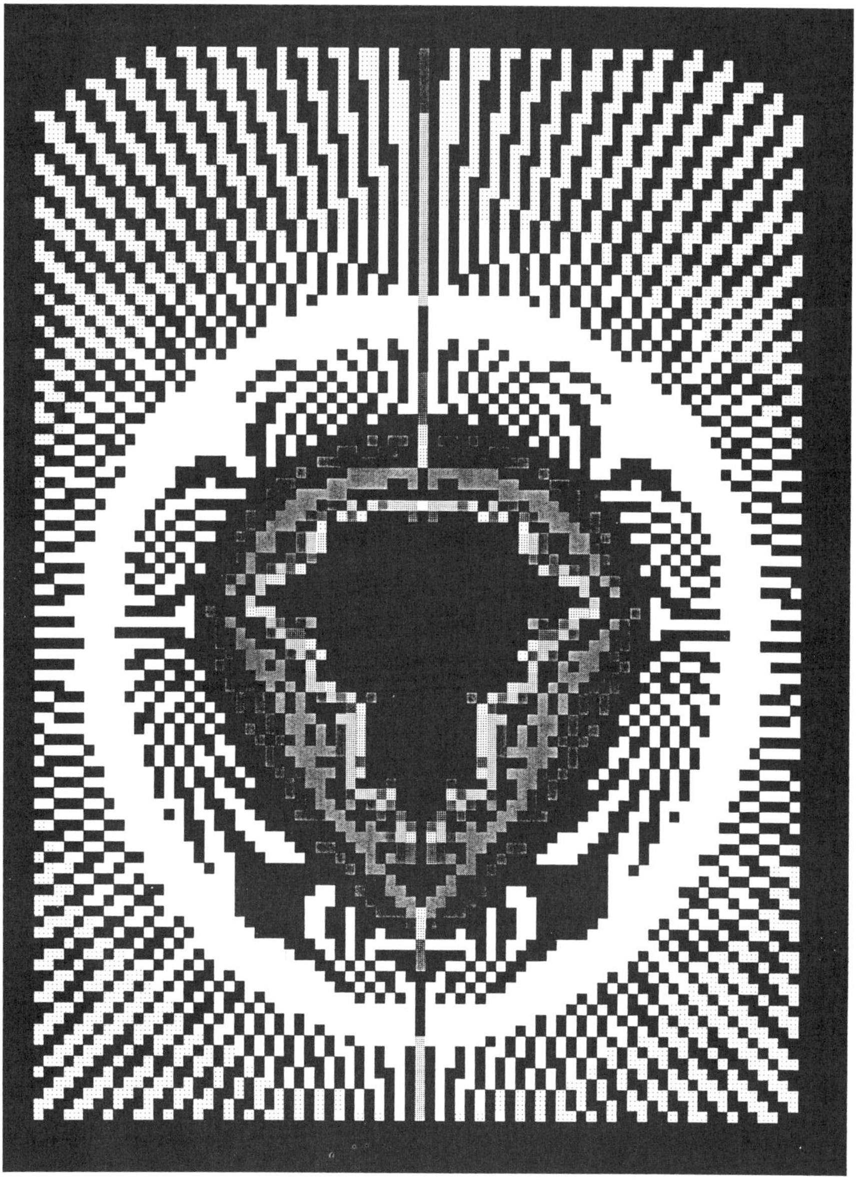

Cycle 77 R = -1.999999999999999999999999999999999999999999999351706492621563711622387677636627 9
781744375528 M = 4.40×10^{90} S = 100x76 D = 1500 AR VAX

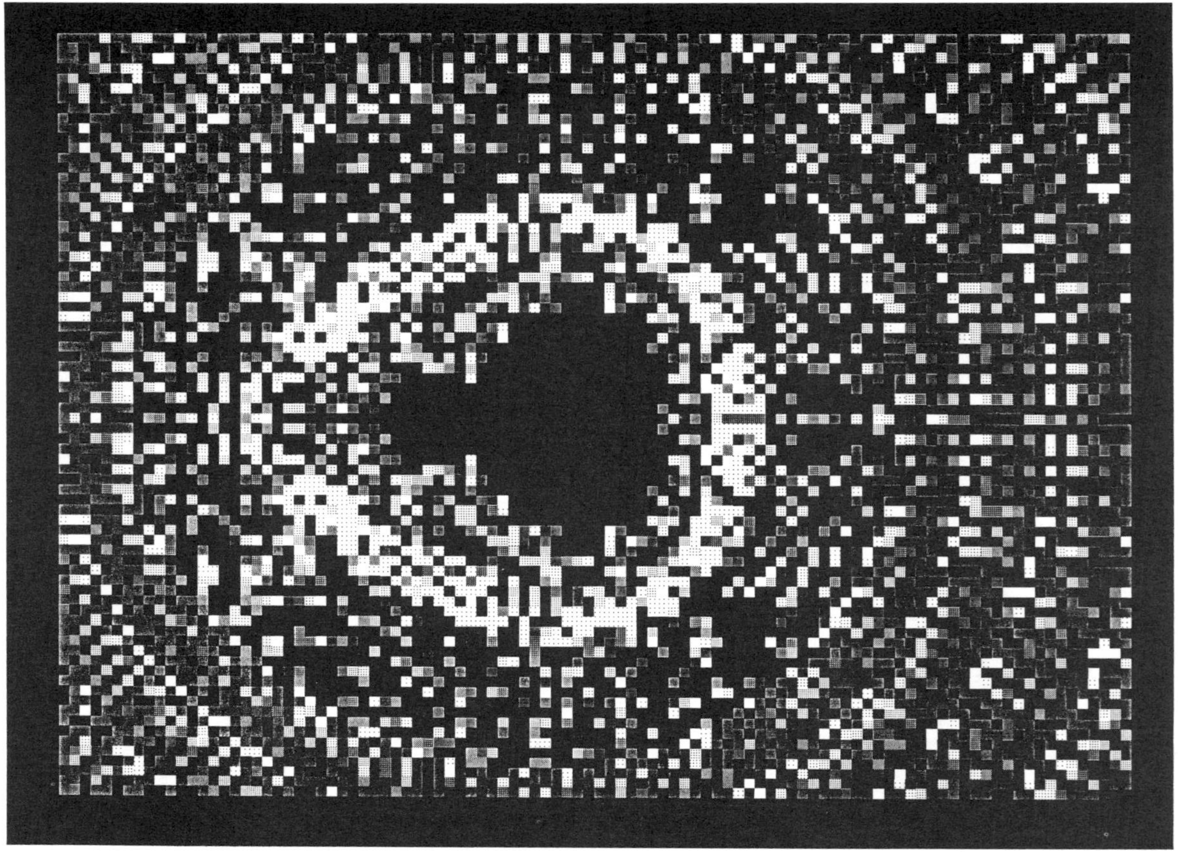

Cycle 78 R = -1.99999999999999999999999999999999999999999999837926623155390927905596919409156 9
9454360938919 M = $7.04x10^{91}$ S = 100x76 D = 1500 AR VAX

Cycle 79 R = -1.99999999999999999999999999999999999995948165578884773197639922985222892 4863590234736OO M = 1.13×10^{93} S = 100x76 D = 1500 AR VAX

Cycle 80 R = -1.99999999999999999999999999999999999999999999998987041394721193299409980746307 23
1215897558684395 M = $1.80 x 10^{94}$ S = 100x76 D = 1500 AR VAX

Cycle 81 R = -1.999999999999999999999999999999999999999974676034868029832485249518657680780397438967112396 M = $2.89x10^{95}$ S = 100x76 D = 2000 AR VAX

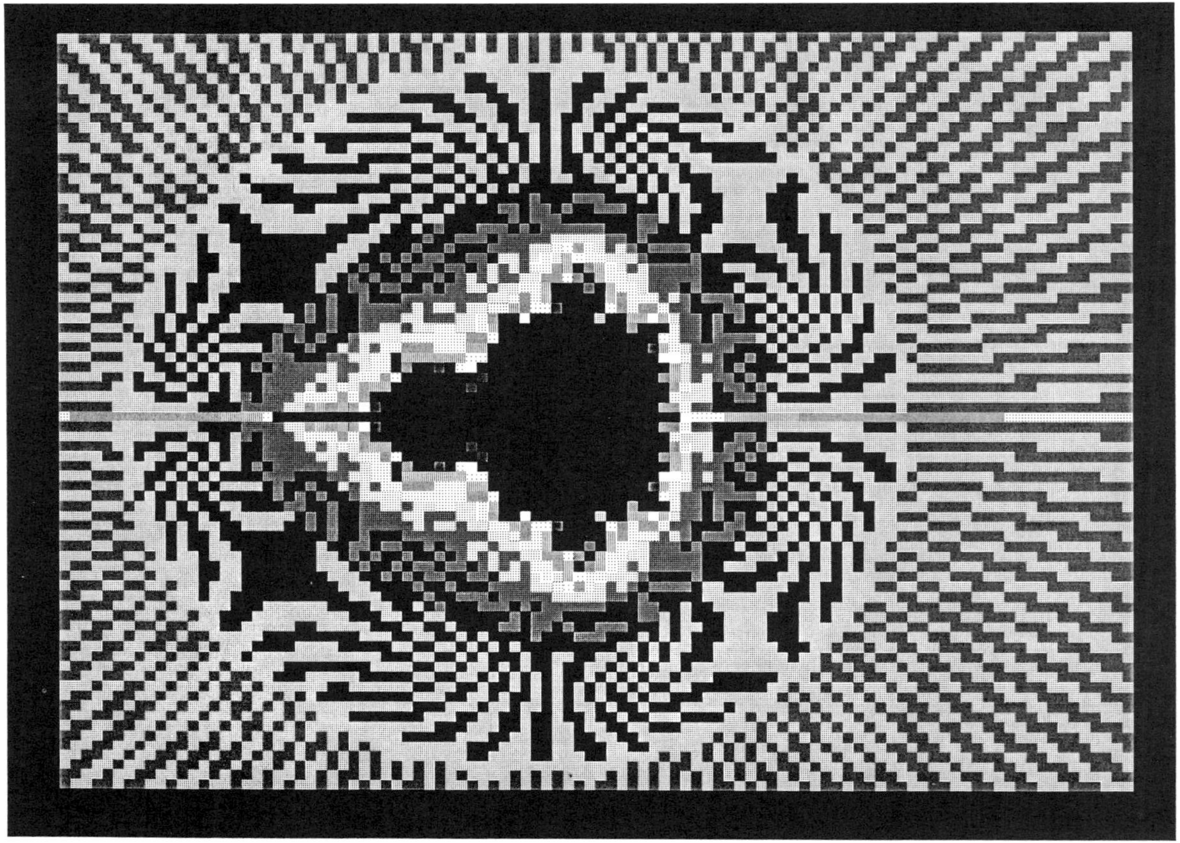

Cycle 82 R = -1.99999999999999999999999999999999999999999999999993669008717007458121312379664420 1950993597417782575 M = 4.62×10^{96} S = 100x76 D = 2000 AR VAX

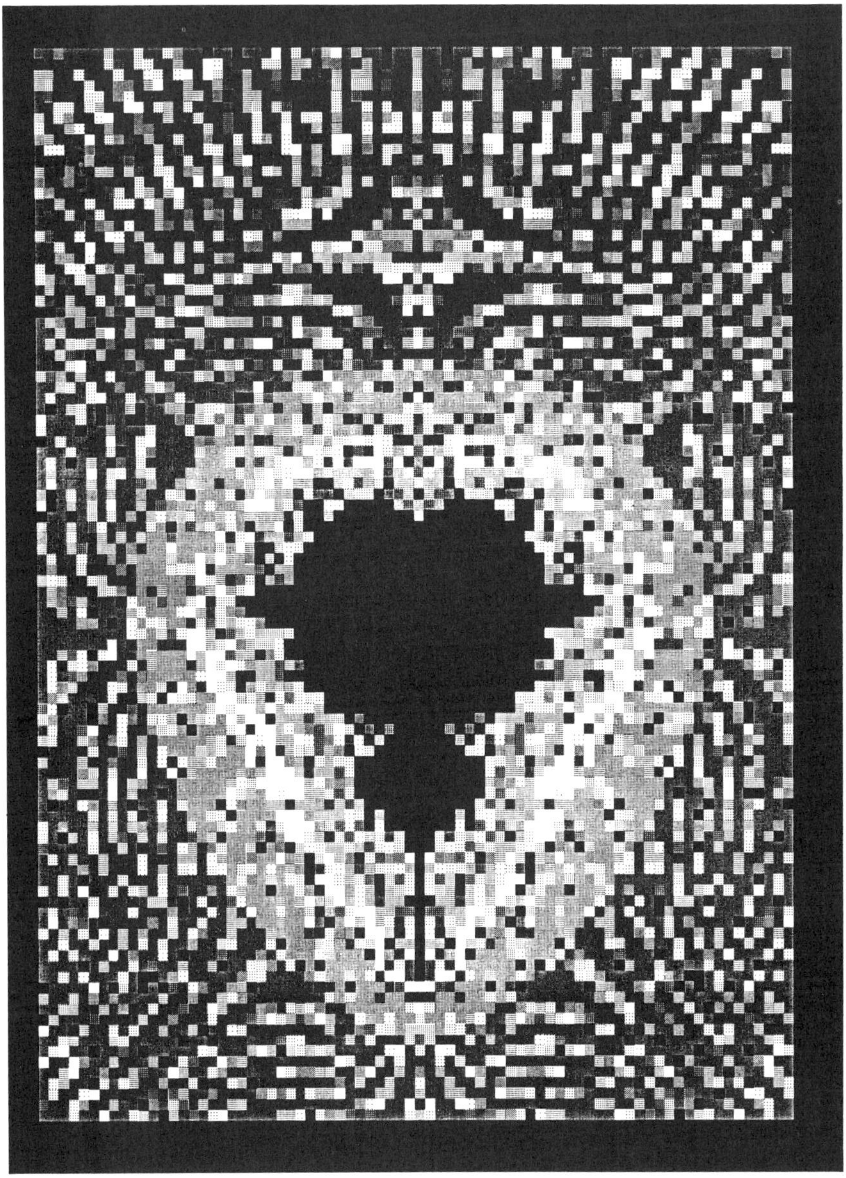

Cycle 83 R = -1.999999999999999999999999999999999999999999999999999999841725217925186453032809491610504877483993544457441 M = 9.24×10^{97} S = 100x76 D = 2500 AGDP ZEN

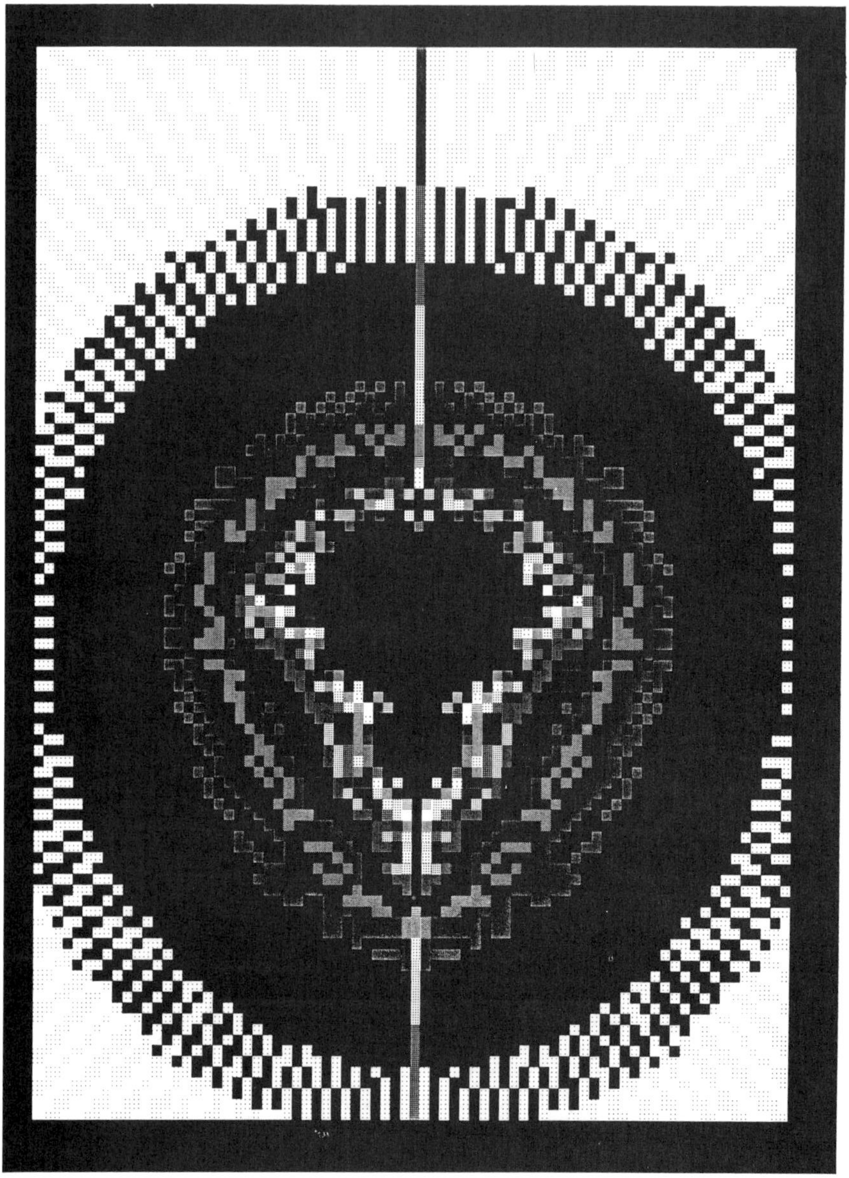

Cycle 84 R = -1.999999999999999999999999999999999999999999999999999960431304481296613258202372902 62621937099838611442374 M = 1.18×10^{99} S = 100x76 D = 2000 AR VAX

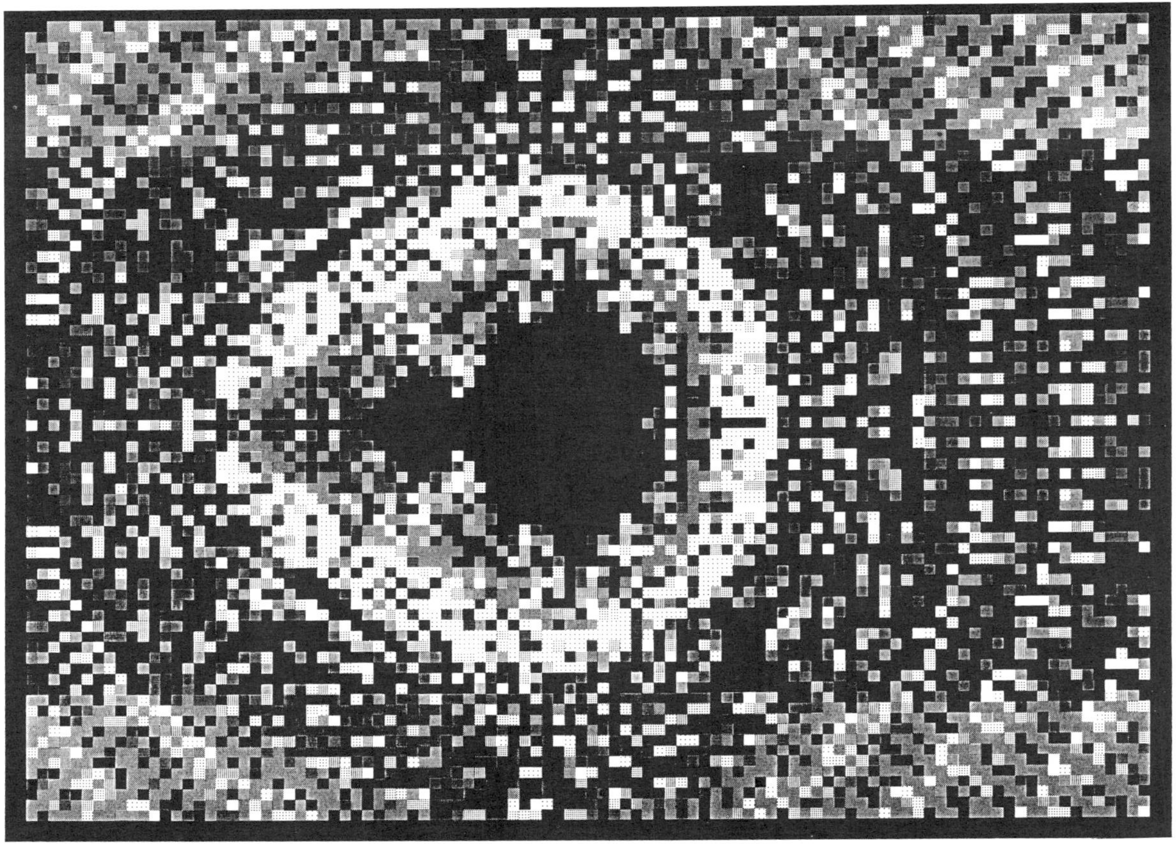

Cycle 85 R = -1.9999999999999999999999999999999999999999999999999901078261203241533145505932256565548427495965286099948 M = 1.89×10^{100} S = 200x151 D = 2000 AR VAX

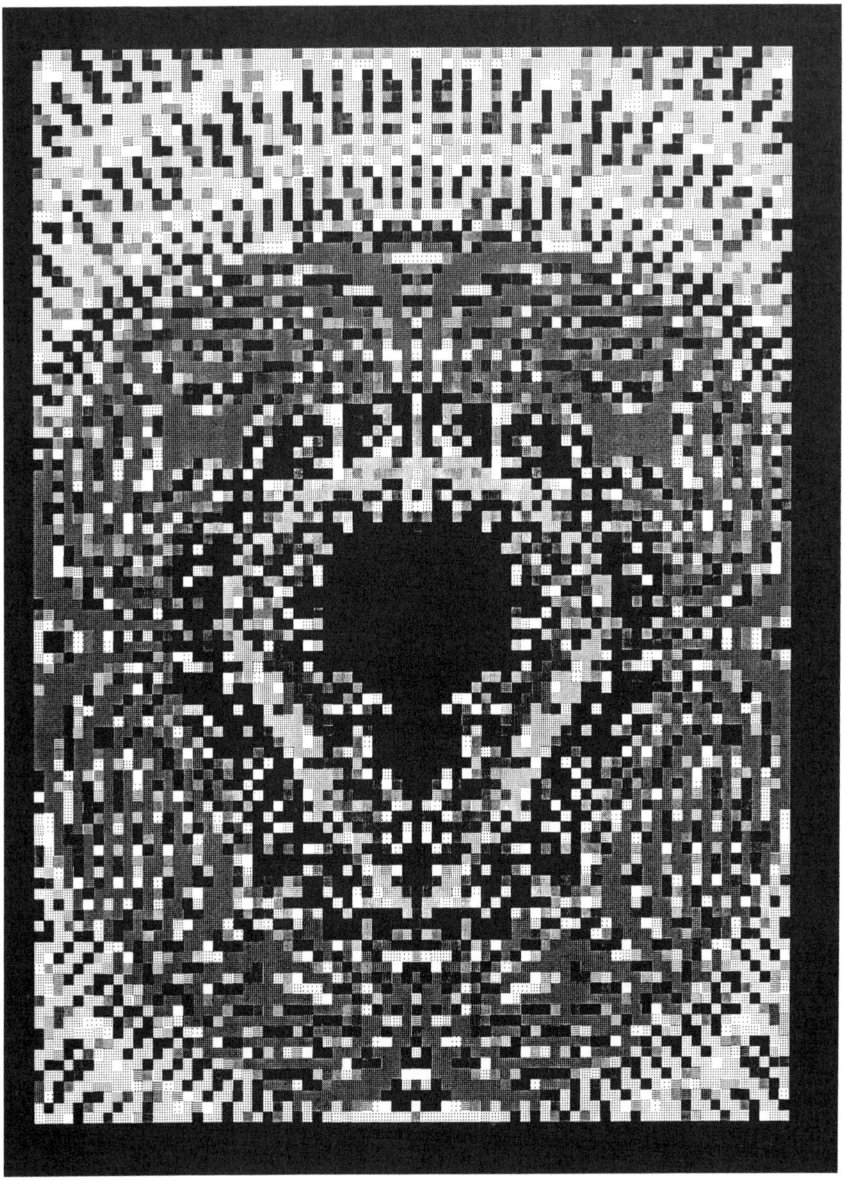

Cycle 95 R = -1.99999999999999999999999999999999999999999999999999999999905660878375283749719148571259084270884986844335663868127676 M = $2.08\text{x}10^{112}$ S = 100x76 D = 3000 AGDP ZEN UC

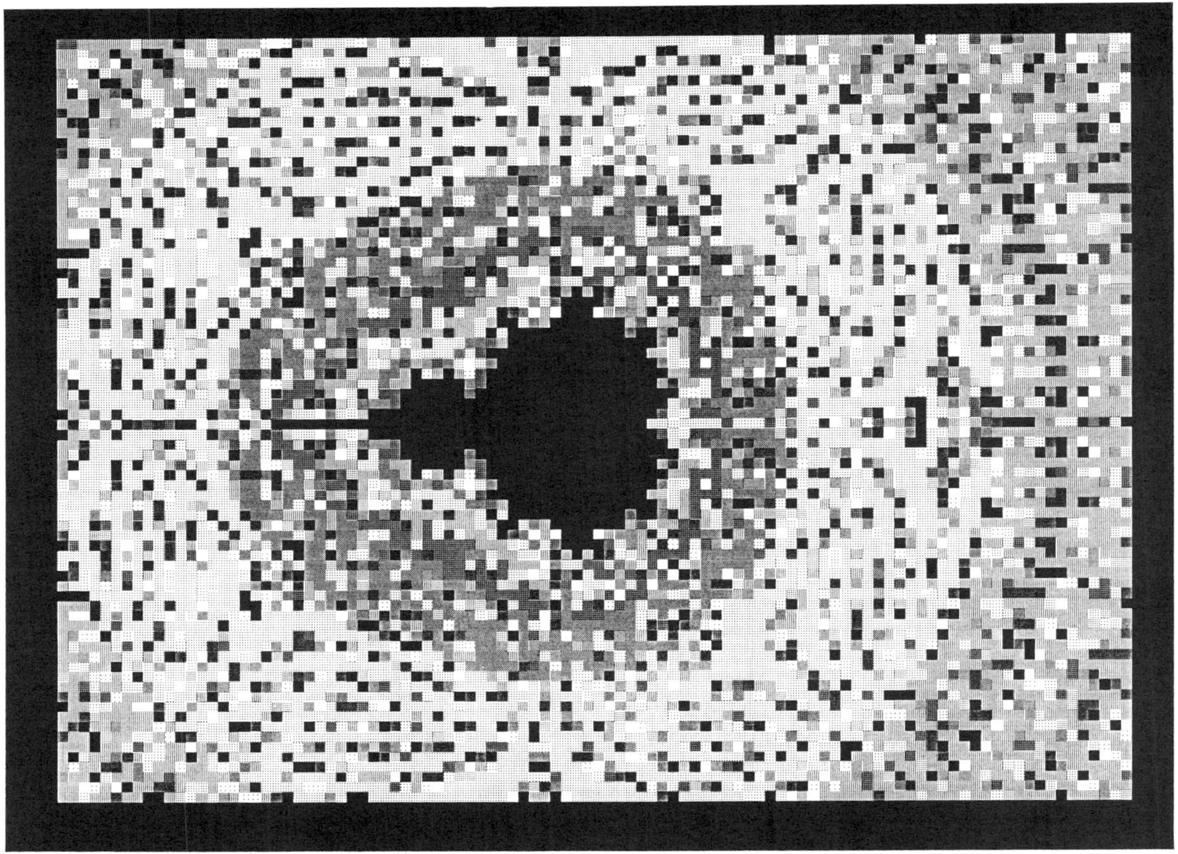

Cycle 100 R = -1.99999999999999999999999999999999999999999999999999999999999078719515383630368351060266201994832861199651715467462319416670 M = 2.18×10^{118} S = 100x76 D = 3000 AGDP ZEN UC

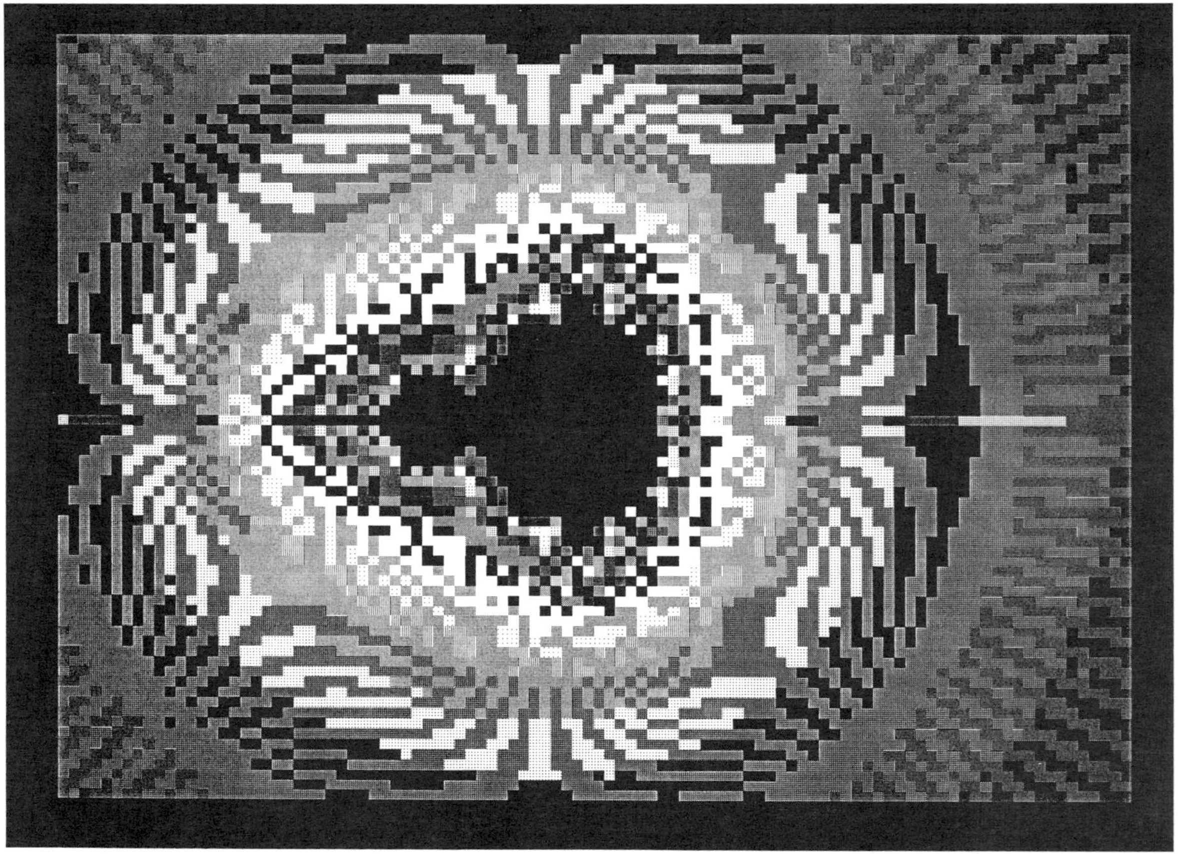

Cycle 105 R = -1.999999999999999999999999999999999999999999999999999999991003120267418265 3159283229121288557896601528487838619367144106110 M = 2.29x10^{124} S = 100x76 D = 3000 AGDP ZEN UC

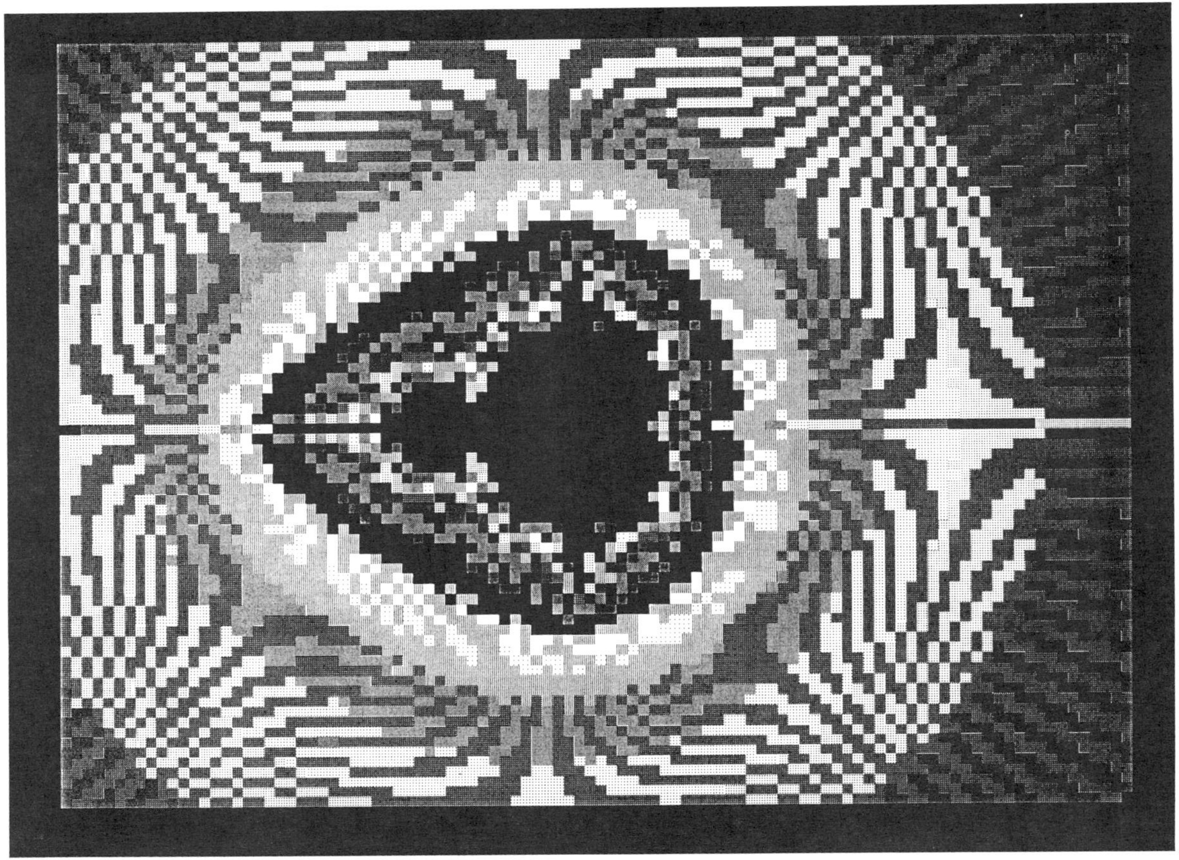

Cycle 110 R = -1.99999999999999999999999999999999999999999999999999999999999999999121398463615064972258625284387583573208999301639049017257266797249 1 M = $2.40x10^{130}$ S = 100x76 D = 3000
AGDP ZEN UC

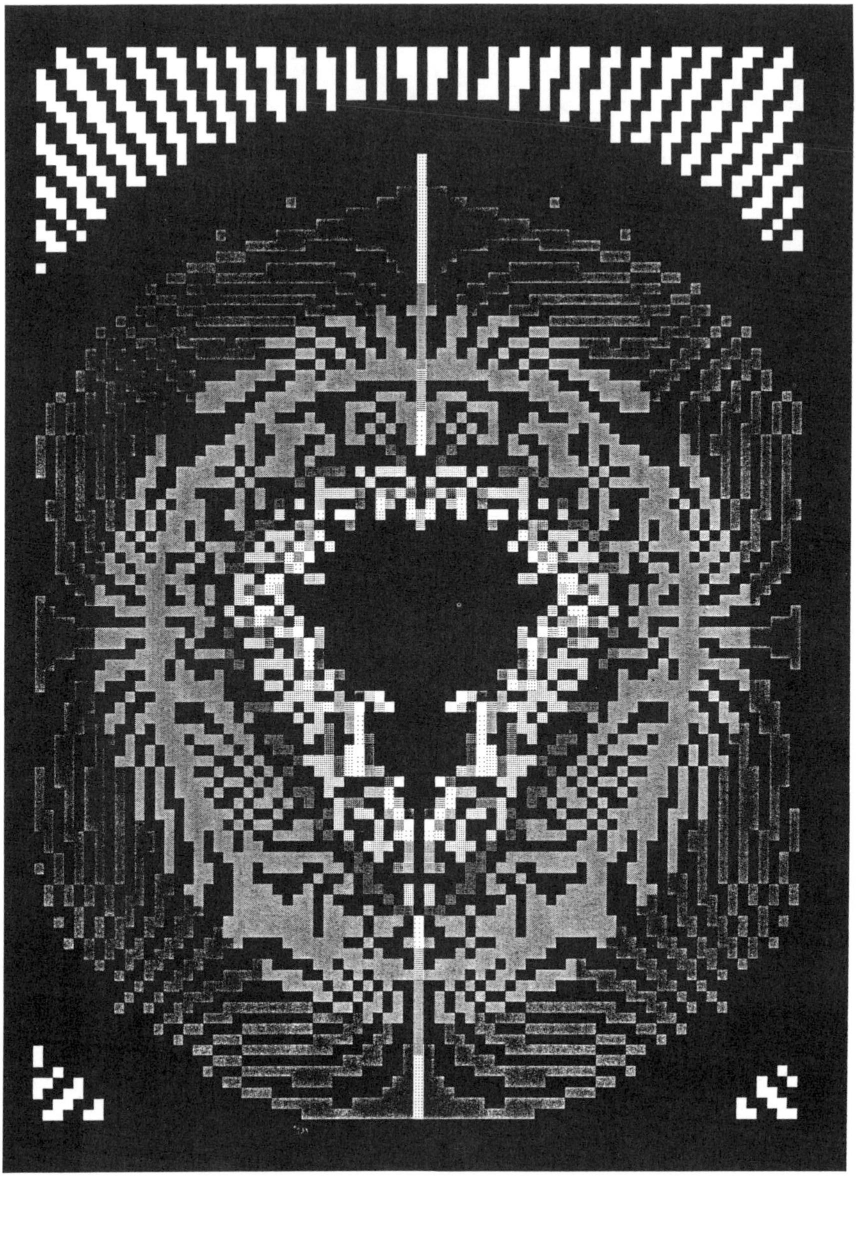

Cycle 115 R = -1.9999999999999999999999999999999999999999999999999999999999999999999999999999991419906871
240868697131375428474958321119133805068838059152996068246260 M = 2.51×10^{136} S = 100x76
D = 3000 AGDP ZEN UC

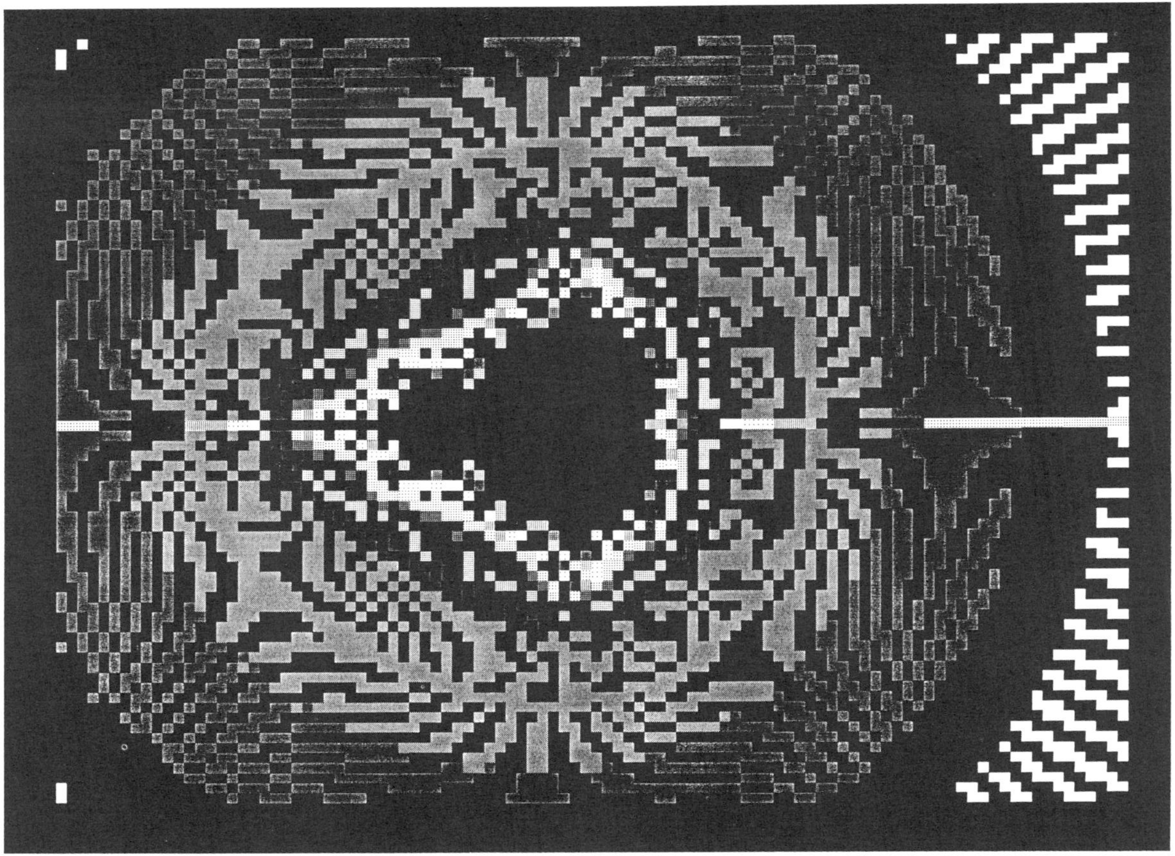

Cycle 120 R = -1.99999999999999999999999999999999999999999999999999999999999999999999991621002 803946161005579235881687007648553841606512537167141597722898095360 M = 2.64x10^{142} S = 100x76 D = 3000 AGDP ZEN + ZEN UC

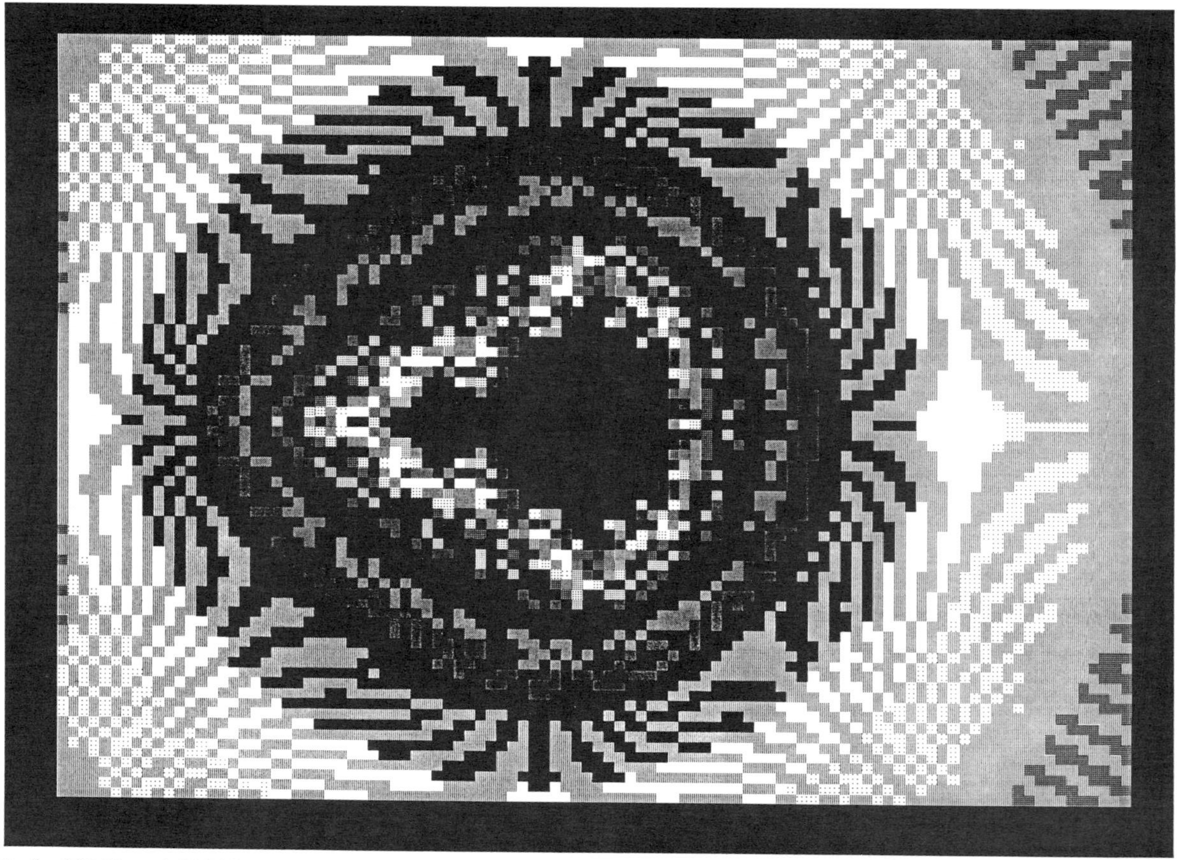

Cycle 125 R = -1.99999999999999999999999999999999999999999999999999999999999999999999999991817
38555072867285701097254070996840679086094385989957728671652626767260 M = 2.67×10^{148} S = 100x76
D = 3000 AGDP ZEN UC

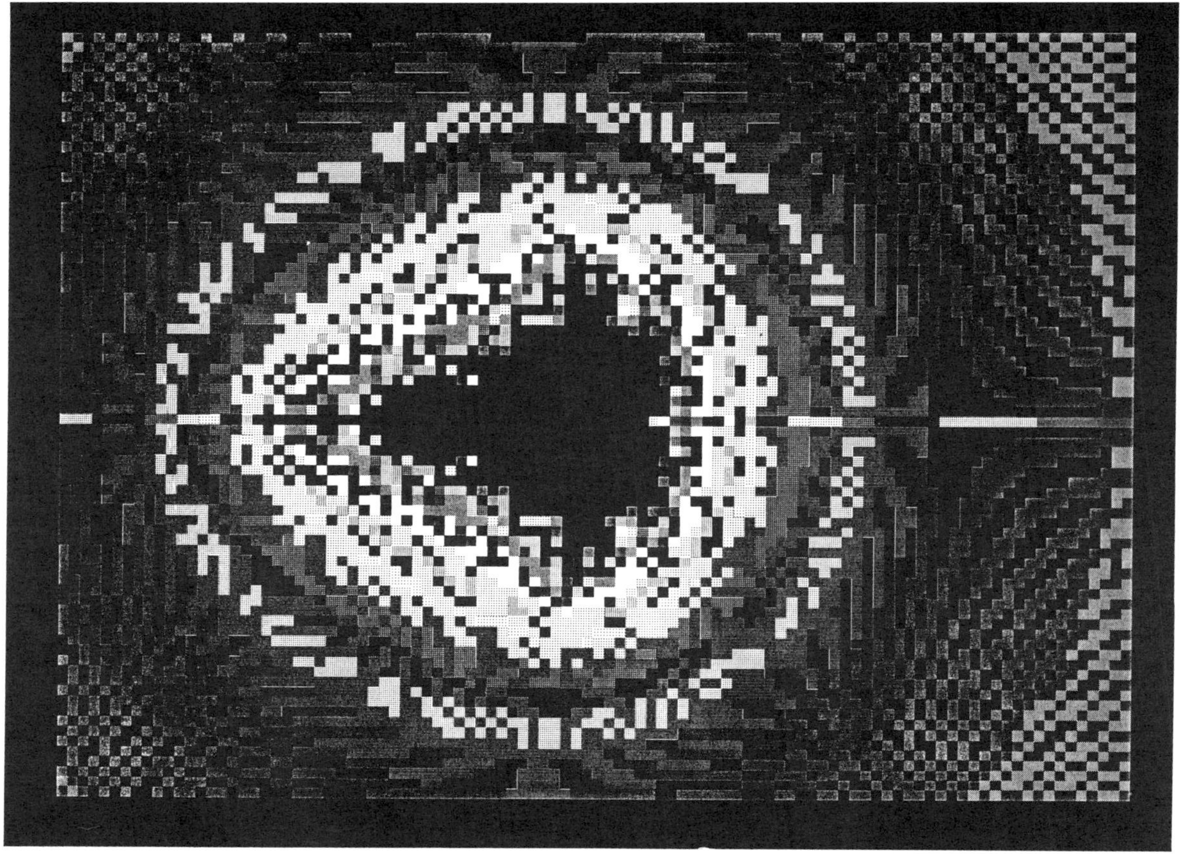

Cycle 150 R = -1.999999999999999999999999999999999999999999999999999999999999999999999999999999
99999999999927323784294307821136724468231360138108872889829291881878307859621447951453738355
735811176795592 M = 1.78×10^{179} S = 100x76 D = 10000 AGDP ZEN UC

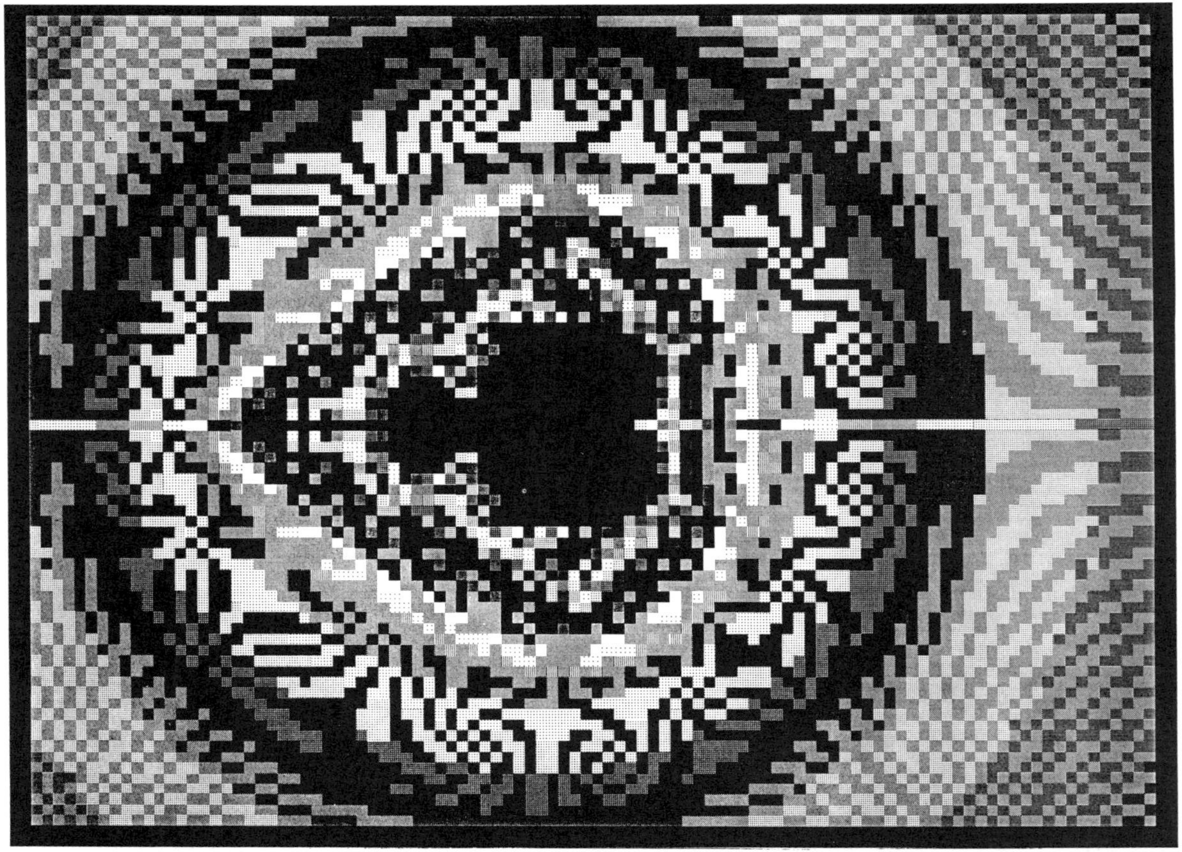

Cycle 175 R = -1.9999999999999999999999999999999999999999999999999999999999999999999999999999999 9999999999999999999999999935450553584732898870090583393164374567228316185644738104006154718 781786042692477188441548689870613445768523494 M = 4.44x10^{208} S = 100x76 D = 10000 PR ZEN + VAX

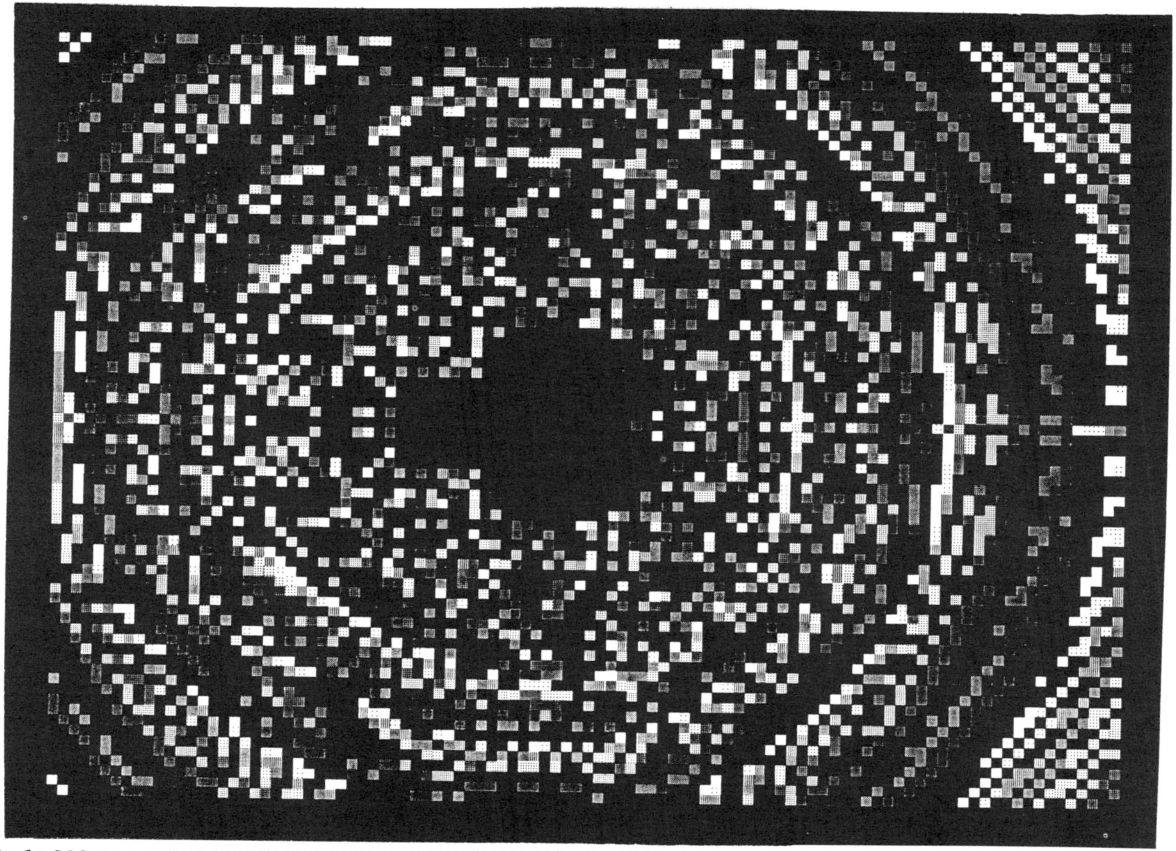

Cycle 200 R = -1.9999999999999999999999999999999999999999999999999999999999999999999999999999999999999999
9999999999999999999999999999999999999999426685746903701524868873414969921982795393910464 9515
64378620882667385903299117419504340886237075855598460105829200658465344388 M = 5.68×10^{238}
S = 100x76 D = 10000 RP DEC SYS5000 + ZEN

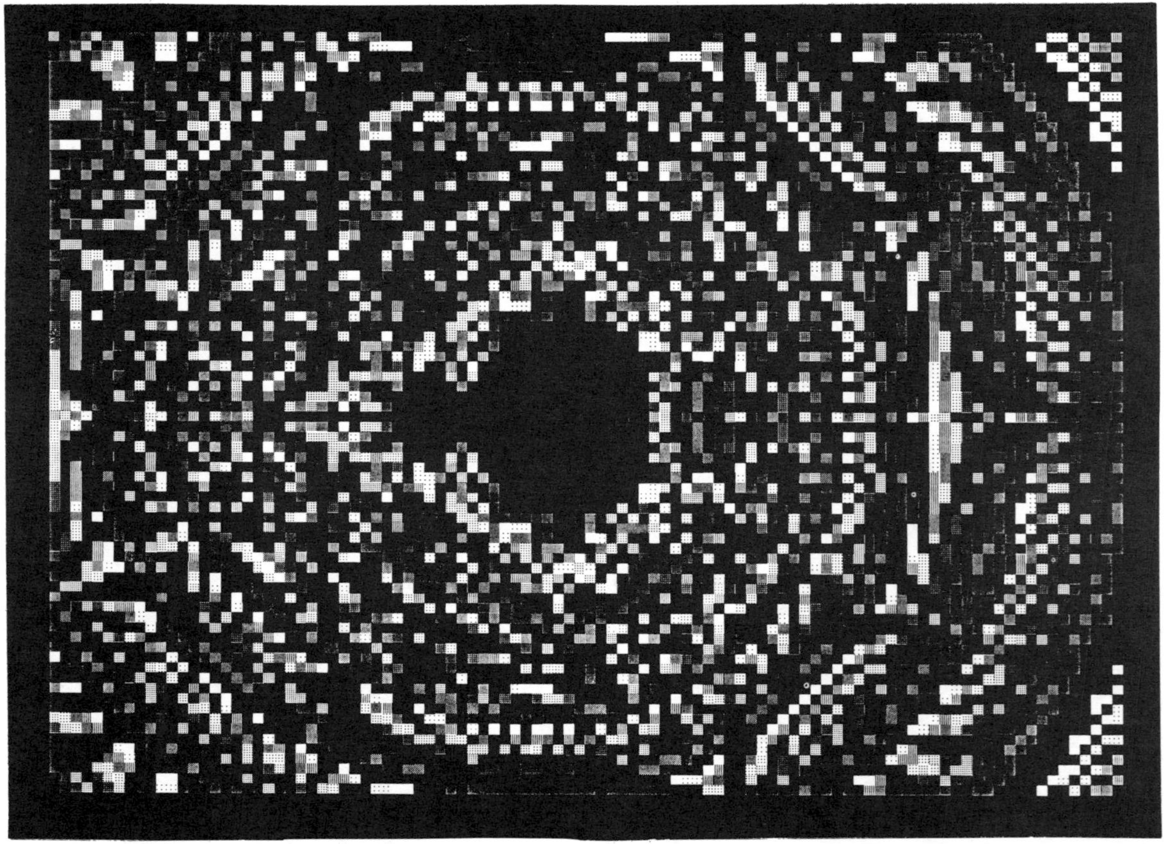

Cycle 250 R = -1.999999999999999999999999999999999999999999999999999999999999999999999999999999 9999999999999999999999999999999999999999999999999999999999999999999999954773479932634571461 28 49868772624150857945748314942293641602039247450498736039839970632632438592848265993465823922 7575629945793894089888138080140518933597742 M = 9.04×10^{298} S = 100x76 D = 10000 RP DEC SYS5000 + ZEN

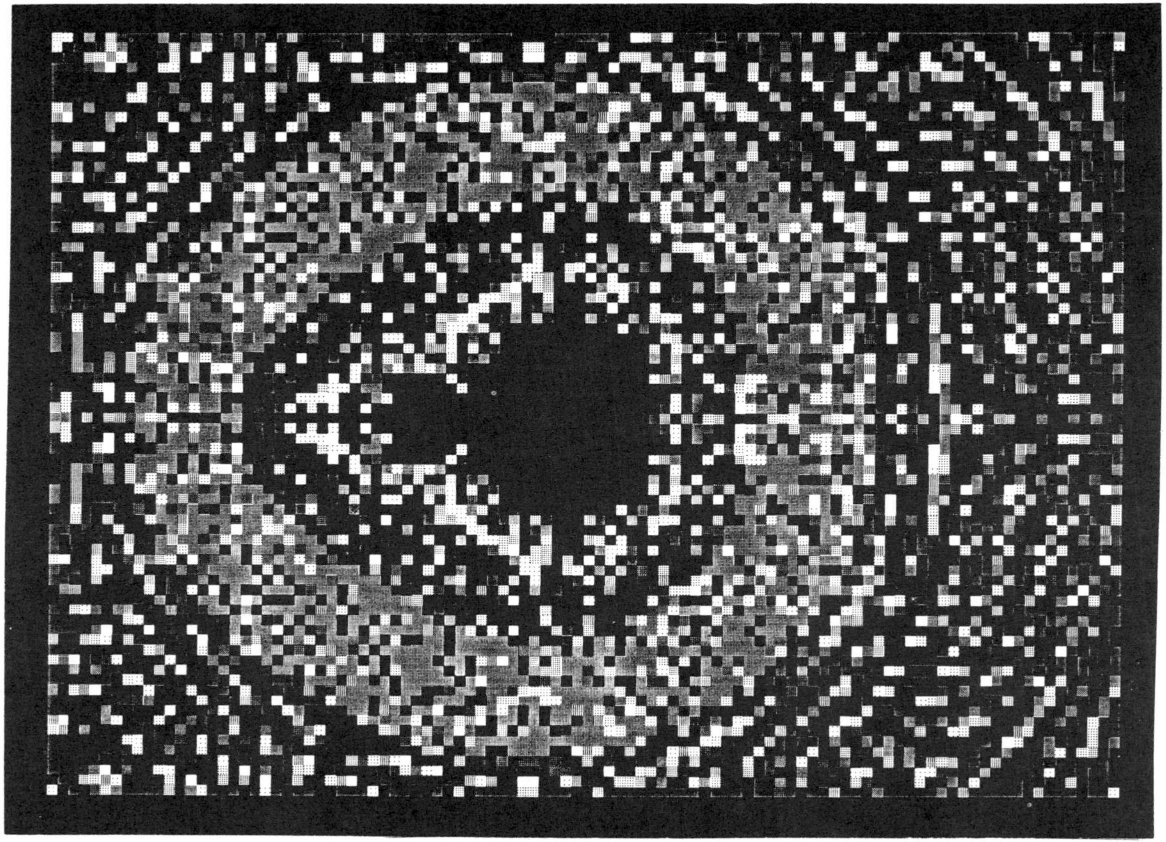

Cycle 300 R = -1.9999999999999999999999999999999999999999999999999999999999999999999999999999999
999999999999999999999999999999999999999999999999999999999999999999999999999999999999999999969
42265581617082564244406089532095341856110462695594154906282867945161787009721229863548534802
541940266148691731543402237271898408295765449289538680351633 M = 1.45×10^{359} D = 10000
AR DEC SYS5000

Eight Midgets at higher resolution

Cycle 28	KWP Mac II
Cycle 39	KWP Mac II
Cycle 44	KWP Mac II
Cycle 49	KWP Mac II
Cycle 57	KWP Mac II
Cycle 60	KWP Mac II
Cycle 70	AGDP Mac II
Cycle 85	AR VAX

Cycle 28 KWP Mac II

Cycle 39 KWP Mac II

Cycle 44 KWP Mac II

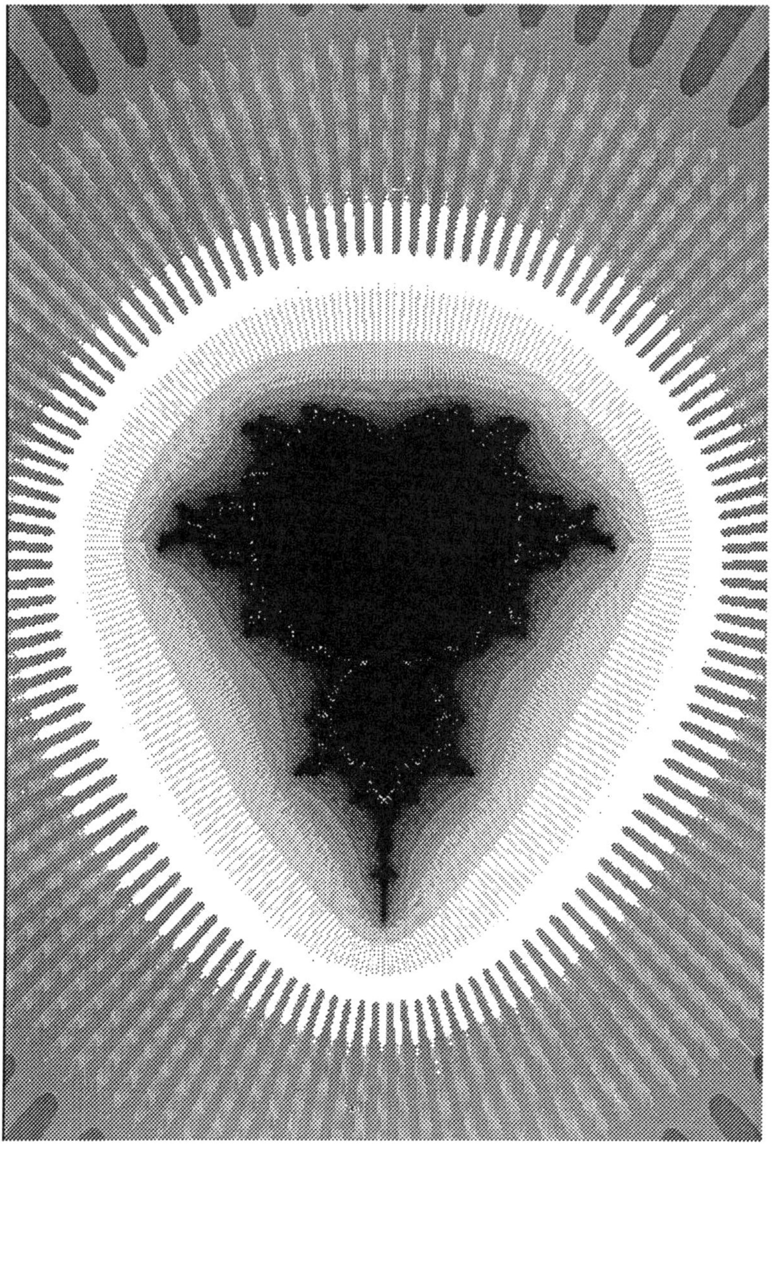

Cycle 49 KWP Mac II

Cycle 57 KWP Mac II

Cycle 60 KWP Mac II

Cycle 70 AGDP Mac II

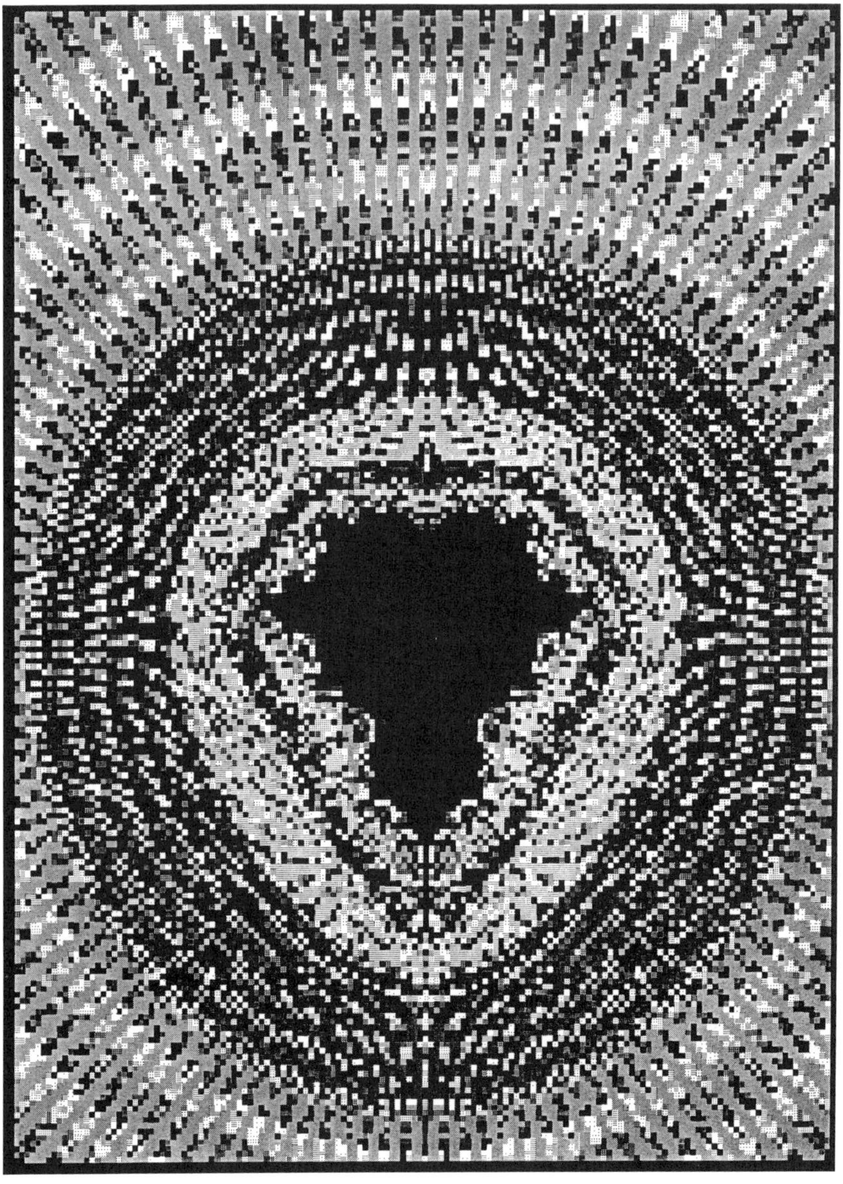

Cycle 85 AR VAX

10.2 APPENDIX II - MATHEMATICAL APPENDIX

To see the structure of K_{-2}, observe the function $g(z) = (z + (z^2 - 4)^{1/2})/2$ takes C-K to C-D, where K is the line segment along the real axis between -2 and 2. Also, observe g(z) has an inverse function $g^{-1}(w) = w + (1/w)$ taking C-D to C-K. Then a tedious calculation shows $g^{-1} \circ F_o \circ g = F_{-2}$ for points outside the unit disk D. Since points outside D diverge to infinity under iteration of F_o, it follows that points outside K diverge to infinity under iteration of F_{-2}. On the other hand, for real x satisfying $-2 \leq x \leq 2$, we see $0 \leq x2 \leq 4$ and so $-2 \leq x^2 -2 \leq 2$. That is, $-2 \leq x \leq 2$ implies $-2 \leq F_{-2}(x) \leq 2$. It follows that the filled-in Julia set $K_{-2} = K$.

Now we begin to sketch how the family of polynomials given by (2) is related to cardioids and disks in the Mandelbrot set. First, consider iteration of functions. The point z_0 is a fixed point of the function $F_c(z)$ if z_0 is a solution of the equation $F_c(z) = z$; that is,

$$z^2 - z + c = 0. \qquad (6)$$

This is a quadratic equation with solutions

$$z_1 = ((1 + (1 - 4c)^{1/2})/2) \text{ and } z_2 = ((1 - (1 - 4c)^{1/2})/2). \qquad (7)$$

We see there are two fixed points as long as $c \neq 1/4$.

A fixed point can be thought of in this way: a single application of the function to the fixed point returns the fixed point. The notion of fixed point can be generalized by allowing the point to be returned only after several applications of the function. These are called periodic points: more specifically, a point z_1 is periodic of period n, or belongs to a cycle n, if there are points $z_2, z_3, .., z_n$ for which $F_c(z_1) = z_2$, $F_c(z_2) = z_3, ..., F_c(z_{n-1}) = z_n$, and $F_c(z_n) = z_1$. The collection of points $\{z_1, z_2, z_3, ..., z_n\}$ is the cycle n. For example, with $c = -1$ the point $z = 0$ is periodic of period two:

$$F_{-1}(0) = -1 \text{ and } F_{-1}(-1) = 0.$$

Functionally, points of a cycle n are solutions of the equation

$$F_c(F_c(\dots (F_c(z)) \dots)) = z,$$

where the left-hand side contains n occurrences of $F_c(z)$. For example, the cycle 2 for equation for $c = -1$ can be written

$$F_{-1}(F_{-1}(0)) = F_{-1}(-1) = 0.$$

Using the notation $F_c^{\circ n}(z)$ for the composition of F_c with itself n times, the cycle n equation can be written

$$F_c^{\circ n}(z) = z \qquad (8)$$

That is, a periodic point of period n is a fixed point for the nth composition of $F_c(z)$.

The cycle n equation (8) is a polynomial equation of degree 2^n - that is, the highest power of z in (8) is 2^n. Equation (6) demonstrates this for fixed points, which can be thought of as cycle 1s. For n = 2 we have

$$z^4 + 2cz^2 - z + (c^2 + c) = 0 \qquad (9)$$

and for n = 3

$$z^8 + 4cz^6 + (6c^2 + 2c)z^4 + (4c^3 + 4c^2)z^2 - z + (c^4 + 2c^3 + c^2 + c) = 0. \qquad (10)$$

Some simplification of these equations is possible. (This is point (ii) of section 4.) For instance, a fixed point z_1 belongs to the cycle 2 $\{z_1, z_1\}$ since it satisfies (8) with n = 2:

$$F_c^{o2}(z_1) = F_c(F_c(z_1)) = F_c(z_1) = z_1.$$

Since every fixed point satisfies (6) and every cycle 2 satisfies (9), fixed points must also satisfy (9) and so the left-hand side of (6) must be a factor of the left-hand side of (9). Dividing these polynomials, we see the cycle 2s which are not fixed points must satisfy

$$z^2 + z + (c + 1) = 0. \qquad (11)$$

The z-values which satisfy (11) constitute the proper cycle 2, for the given value of c. Similarly, proper cycle 3s must satisfy

$$z^6 + z^5 + (3c + 1)z^4 + (2c + 1)z^3 + (3c^2 + 3c + 1)z^2 + (c^2 + 2c + 1)z + (c^3 + 2c^2 + c + 1) = 0.$$

For higher-order cycles, other simplifications are possible. For instance, every cycle 2 is also a cycle 4. These observations must be kept in mind when we locate cycles of a particular order.

Usually, periodic points are numerically invisible unless they are stable. A fixed point z is stable if all points near enough to z will converge to z under iteration of F_c. That is, there is some number for which any point w a distance less than from z satisfies $F^{on}(w) \rightarrow z$ as $n \rightarrow \infty$. There is an analogous notion of stability for periodic points. For instance, the cycle 2 $\{z_1, z_2\}$ is stable if for any point w near enough to z_1, $F_c^{o2n}(w) \rightarrow z$ and $F_c^{o2n+1}(w) \rightarrow z_2$. Since F_c is continuous, the same condition automatically holds for points near z_2. Alternately, the fixed point stability condition can be applied to the composition $F^{o2}(z)$.

Finding these neighborhoods from which points give rise to sequences converging to periodic points, may be a difficult task. Fortunately, for differentiable functions there is a direct method for determining if a periodic point is stable. A fixed point z_1 is stable if the derivative $F_c'(z_1)$ satisfies

$$|F_c'(z_1)| < 1.$$

This stability criterion holds for fixed points of any differentiable function; it is especially simple for $F_c(z)$ since $F_c'(z) = 2z$. That is, a fixed point z_1 is stable if

$$2|z_1| < 1. \tag{12}$$

Using the chain rule for differentiation, the stability of periodic points can be checked in this way. The cycle 2 $\{z_1, z_2\}$ is stable if

$$1 > |(F_c^{o2})'(z_1)| = |F_c'(F_c(z_1))F_c'(z_1)|$$

$$= |F_c'(z_2)F_c'(z_1)| = |F_c'(z_2)||F_c'(z_1)|,$$

that is

$$1 > |F_c'(z_2)||F_c'(z_1)| = |(2z_1)(2z_2)| = 2^2|z_1z_2|. \tag{13}$$

There is nothing special about the choice of z_1 as the starting point, since $|(F_c^{o2})'(z_2)| = |F_c'(z_1)||F_c'(z_2)|$. This generalizes in the obvious way to higher-order cycles.

To approach the problem of finding stable periodic points of $F_c(z)$ we use a theorem of Fatou: for any stable cycle n of a polynomial $g(z)$, there is a critical point z_0 of $g(z)$ (i.e., $g'(z_0) = 0$) which under iteration of g converges to the cycle n. More briefly, every stable cycle n of a polynomial attracts a critical point of the polynomial. For $F_c(z) = z^2 + c$ there is only one critical point, $z = 0$, and so for each value of c there is at most one stable cycle n.

We have seen already how to test stability of fixed points and cycle-2s - (12) and (13). In general, an cycle n $\{z_1, z_2, \ldots, z_n\}$ is stable if

$$|F_c'(z_n)||F_c'(z_{n-1})|\ldots|F_c'(z_1)| < 1, \tag{14}$$

or using the value of the derivative $F_c'(z) = 2z$, we see the cycle n is stable if

$$2^n|z_n||z_{n-1}|\ldots|z_1| < 1. \tag{15}$$

Since every polynomial equation can be solved with complex numbers, for every c there are solutions of $F_c^{on}(z) = z$ - that is, there are cycle ns. However, the stability of these cycle ns is determined by (15), and by Fatou's theorem we know for each c at most one of these cycles is stable. We shall work through two examples, stable fixed point and stable cycle 2, which provide an observation we need for finding midgets.

The two fixed points are given by (7); these are stable when

$$|1 + (1 - 4c)^{1/2}| < 1 \text{ and when } |1 - (1 - 4c)^{1/2}| < 1, \text{ respectively.} \tag{16}$$

We must solve these equations to find the c values for which there is a stable fixed point. Solving these equations involves taking square roots of complex numbers, which is most easily accomplished through the polar co-ordinate representation. So write 1 - 4c as $r\cos(\theta) + r\sin(\theta)i$. Then

$$(1 - 4c)^{1/2} = r^{1/2}\cos(\theta/2) + r^{1/2}\sin(\theta/2)i$$

and so (16) becomes

$$((1 + r^{1/2}\cos(\theta/2))^2 + (r^{1/2}\sin(\theta/2)i)^2)^{1/2} < 1$$

and

$$((1 - r^{1/2}\cos(\theta/2))^2 + (r^{1/2}\sin(\theta/2)i)^2)^{1/2} < 1.$$

Taking square roots of complex numbers involves some delicate issues, in the polar representation we take $-\pi \le \theta < \pi$. Consequently, $\cos(\theta/2) \geq 0$ and so

$$(1 - r^{1/2}\cos(\theta/2))^2 \leq (1 + r^{1/2}\cos(\theta/2))^2,$$

whence $|z_1| \geq |z_2|$. Since Fatou's theorem implies for any value of c there can be at most one stable fixed point; z_2 is the stable fixed point if there is one.

To find the boundary of the region of c-values for which there is a stable fixed point, we solve $|2z_2| = 1$, that is,

$$|(1 - r^{1/2}\cos(\theta/2)) + (r^{1/2}\sin(\theta/2)i)| = 1.$$

Squaring both sides yields

$$(1 - r^{1/2}\cos(\theta/2))^2 + (r^{1/2}\sin(\theta/2))^2 = 1$$

and solving for r

$$r = 2(1 + \cos(\theta)). \qquad (17)$$

(Some trigonometry is required to get to (17): we recall $\cos^2(\theta/2) + \sin^2(\theta/2) = 1$ and $\cos^2(\theta/2) = +((1 + \cos(\theta))/2)^{1/2}$.) The relation (17) describes a cardioid with cusp at the origin and farthest right extent 4. Recalling this gives the values, not of c, but of 1 - 4c, we can retrieve the c values for which there is a stable fixed point through the following calculations:

$$1 - 4c = r\cos(\theta) + r\sin(\theta)i = 2(1+\cos(\theta))\cos(\theta) + 2(1+\cos(\theta))\sin(\theta)i$$

and so writing $c = x + yi$ we find

$$x = (1/4) - (1/4)(2 + 2\cos(\theta))\cos(\theta)$$

$$y = -(1/4)(2 + 2\cos(\theta))\sin(\theta)$$

This is the main cardioid of the Mandelbrot set: for any c inside this cardioid, $F_c(z)$ has exactly one stable fixed point.

From (11) we see the points of the cycle 2 are given by

$$z_1 = ((-1 + (-3 - 4c)^{1/2})/2) \text{ and } z_2 = ((-1 - (-3 + 4c)^{1/2})/2)$$

and from (13) stability is determined by the condition $4|z_1z_2| < 1$. Substituting in the values of z_1 and z_2 yields $|4(c+1)| < 1$; that is, the c values for which there is a stable cycle 2 form a disk of radius 1/4 and centered at $c = -1$. This is called the head of the Mandelbrot set: for any c inside this disk, $F_c(z)$ has a stable cycle 2.

This open cardioid and open disk are hyperbolic components of the Mandelbrot set. The hyperbolic components are maximal open pieces of the Mandelbrot set to which there corresponds a stable cycle n (constant n throughout the piece) for each c in the component. For each c in the cardioid there is a stable fixed point; for each c in the head there is a stable cycle 2. For larger n there are several components - for instance, there are three hyperbolic components with stable cycle 3s. These are the cycle 3 disks (north and south radicals) attached to the top and bottom of the main cardioid, and the main cardioid of the largest midget along the Spike. This is the reason for calling this midget the cycle 3 midget.

For each hyperbolic component there is a function (basically the left side of (15)) taking c to the product of the derivative of F_c evaluated at each point of the stable cycle. This function is called the multiplier of the hyperbolic component. Taking a point c in the component to the magnitude of the multiplier defines a function from the hyperbolic component to the open unit disk. Since this function is known to be 1-1, only one point of the hyperbolic component corresponds to the center of the disk, so this point is called the center of that component. That is, if c is the center of an cycle n hyperbolic component, the multiplier at c is 0. From (15), we see this means some point of the cycle must itself be 0, so 0 is a fixed point of F_c^{on}. That is, the center of an cycle n hyperbolic component is a solution of

$$F_c^{on}(0) = 0$$

or

$$F_c^{on-1}(c) = 0 \qquad (18)$$

The left-hand sides of these equations give the polynomials of (2), with the convention that $F_c^{o0}(c) = c$.

A derivation of some useful properties of these polynomials can be found in (Frame, Philip and Robucci 1991).

10.3 APPENDIX III - TABLE OF POSITIONS

Cycle # Real Pos. Magnification Resolution Dwell Computer Sys.

Cycle 3 R = -1.7548 M = $2.78x10^{1}$ S = 640x480 D = 1000 AR FRA12 AT
Cycle 4 R = -1.94079 M = $4.42x10^{2}$ S = 640x480 D = 1000 AR FRA12 AT
Cycle 5 R = -1.985424 M = $8.62x10^{3}$ S = 640x480 D = 1000 AR FRA12 AT
Cycle 6 R = -1.9963761 M = $1.40x10^{5}$ S = 640x480 D = 1000 AR FRA12 AT
Cycle 7 R = -1.99909568 M = $2.26x10^{6}$ S = 640x480 D = 1000 AR FRA12 AT
Cycle 8 R = -1.999774048 M = $3.62x10^{7}$ S = 640x480 D = 1000 AR FRA12 AT
Cycle 9 R = -1.9999435217 M = $5.81x10^{8}$ S = 640x480 D = 1000 AR FRA12 AT
Cycle 10 R = -1.9999858811 M = $9.28x10^{9}$ S = 640x480 D = 1000 AR FRA12 AT
Cycle 11 R = -1.9999964703350 M = $1.49x10^{11}$ S = 640x480 D = 1000 AR FRA12 AT
Cycle 12 R = -1.999999117587260 M = $2.35x10^{12}$ S = 640x480 D = 1000 AR FRA12 AT
Cycle 13 R = -1.9999997793970588 M = $3.04x10^{13}$ S = 400x300 D = 1500 AR VAX
Cycle 14 R = -1.99999994484928145 M = $4.87x10^{14}$ S = 400x300 D = 1500 AR VAX
Cycle 15 R = -1.999999986212321505 M = $7.79x10^{15}$ S = 200x150 D = 1000 AR VAX
Cycle 16 R = -1.9999999965530804536 M = $1.24x10^{17}$ S = 200x150 D = 1000 AR VAX
Cycle 17 R = -1.99999999913827011860 M = $1.99x10^{18}$ S = 200x150 D = 1000 AR VAX
Cycle 18 R = -1.999999999784567530000 M = $3.19x10^{19}$ S = 200x150 D = 1000 AR VAX
Cycle 19 R = -1.9999999999461418825233 M = $5.104x10^{20}$ S = 200x150 D = 1000 AR VAX
Cycle 20 R = -1.999999999986535470632367 M = $1.02x10^{22}$ S = 200x150 D = 1000 AR VAX
Cycle 21 R = -1.9999999999966338676581938 M = $1.31x10^{23}$ S = 200x150 D = 1000
AR VAX
Cycle 22 R = -1.999999999999158466914555203 M = $2.09x10^{24}$ S = 200x150 D = 1000
AR VAX
Cycle 23 R = -1.99999999999978961672863924379 M = $3.34x10^{25}$ S = 200x150 D = 1000
AR VAX

Cycle 24 R = -1.999999999999947404182159840012 M = $5.35x10^{26}$ S = 200x150 D = 1000
AR VAX
Cycle 25 R = -1.9999999999999868510455399619060 M = $8.56x10^{27}$ S = 200x150
D = 1000 AR VAX
Cycle 26 R = -1.99999999999999967127613849906008 M = $1.35x10^{29}$ S = 200x150
D = 1000 AR VAX
Cycle 27 R = -1.999999999999999178190346247658319 M = $2.19x10^{30}$ S = 200x150
D = 1000 AR VAX
Cycle 28 R = -1.9999999999999997945475865619151078 M = $3.51x10^{31}$ S = 400x300
D = 1500 AR VAX
Cycle 29 R = -1.99999999999999994863689664047881127 M = $3.51x10^{32}$ S = 200x150
D = 1500 AR VAX
Cycle 30 R = -1.999999999999999987159224160119705045 M = $5.62x10^{33}$ S = 200x150
D = 1500 AR VAX
Cycle 31 R = -1.9999999999999999967898060400299264058 M = $1.56x10^{35}$ S = 100x75
D = 1000 AR VAX
Cycle 32 R = -1.99999999999999999919745151000748161079 M = $4.69x10^{36}$ S = 100x75
D = 1000 AR VAX
Cycle 33 R = -1.999999999999999999799362877501870403303 M = $4.60x10^{37}$ S = 100x75
D = 1000 AR VAX
Cycle 34 R = -1.99999999999999999994984071937546760086492 M = $1.56x10^{39}$
S = 100x75 D = 1000 AR VAX
Cycle 35 R = -1.999999999999999999987460179843866900218748 M = $2.00x10^{40}$
S = 100x75 D = 1000 AR VAX
Cycle 36 R = -1.99999999999999999999686504496096672505484931 M = $4.30x10^{41}$
S = 100x75 D = 1000 AR VAX
Cycle 37 R = -1.999999999999999999999216261240241681263722775 M = $2.50x10^{42}$
S = 100x75 D = 1000 AR AT
Cycle 38 R = -1.9999999999999999999998040653100604203159313660 M = $5.88x10^{43}$
S = 200x150 D = 1000 AR AT
Cycle 39 R = -1.99999999999999999999995101632751510507898288471 M = $7.06x10^{44}$
S = 100x75 D = 1000 AR AT

Cycle 40 R = -1.999999999999999999999987754081878776269745723952
M = 9.97×10^{45} S = 100x75 D = 1000 AR AT
Cycle 41 R = -1.9999999999999999999999969385204696940674364311663
M = 1.41×10^{47} S = 100x75 D = 1000 AGDP ZEN
Cycle 42 R = -1.99999999999999999999999923463011742351685910780301
M = 3.16×10^{48} S = 100x75 D = 1000 AR AT
Cycle 43 R = -1.999999999999999999999999808657529355879214776951487
M = 5.05×10^{49} S = 100x75 D = 1000 AGDP ZEN
Cycle 44 R = -1.9999999999999999999999999521643823389698036942379187
M = 8.09×10^{50} S = 100x75 D = 1000 AR AT
Cycle 45 R = -1.999999999999999999999999988041095584742450923559482681
M = 1.29×10^{52} S = 100x75 D = 1000 AGDP ZEN
Cycle 46 R = -1.9999999999999999999999999970102738961856127308898708625
M = 2.07×10^{53} S = 100x75 D = 1000 AR AT
Cycle 47 R = -1.99999999999999999999999999925256847404640318272246772793
M = 3.31×10^{54} S = 100x76 D = 1000 AGDP ZEN
Cycle 48 R = -1.999999999999999999999999999813142118511600795680616932770
M = 5.30×10^{55} S = 100x76 D = 1000 AGDP ZEN
Cycle 49 R = -1.9999999999999999999999999999532855296279001989201542332428
M = 8.48×10^{56} S = 100x76 D = 1000 AGDP ZEN
Cycle 50 R = -1.999999999999999999999999999988321382406975049730038558313920
M = 1.36×10^{58} S = 100x76 D = 1000 AGDP ZEN
Cycle 51 R = -1.9999999999999999999999999999970803456017437624325096395786848
M = 2.17×10^{59} S = 100x76 D = 1000 AGDP ZEN
Cycle 52 R = -1.99999999999999999999999999999927008640043594060812740989468427
M = 3.47×10^{60} S = 100x76 D = 1000 AGDP ZEN
Cycle 53 R = -1.999999999999999999999999999999817521600108985152031852473671900
M = 5.56×10^{61} S = 100x 76 D = 1000 AGDP ZEN
Cycle 54 R = -1.999999999999999999999999999999954380400027246288007963118418028
1 M = 7.11×10^{62} S = 100x76 D = 1000 AGDP ZEN
Cycle 55 R = -1.999999999999999999999999999999988595100006811572001990779604510
41 M = 1.14×10^{64} S = 100x76 D = 1000 AGDP ZEN

Cycle 56 R = -1.999999999999999999999999999999999997148775001702893000497694901127819 M = 1.82x10^{65} S = 100x76 D = 1000 AGDP ZEN

Cycle 57 R = -1.9999999999999999999999999999999999992871937504257232501244237252819684 M = 2.91x10^{66} S = 100x76 D = 1000 AGDP ZEN

Cycle 58 R = -1.99999999999999999999999999999999999982179843760643081253110593132049299 M = 4.66x10^{67} S = 100x76 D = 1000 AGDP ZEN

Cycle 59 R = -1.9999999999999999999999999999999999999554496094016077031327764828 30123304 M = 7.46x10^{68} S = 100x76 D = 1500 AGDP ZEN

Cycle 60 R = -1.99999999999999999999999999999999999998886240235040192578319412070753082974 M = 1.19x10^{70} S = 100x76 D = 1500 AGDP ZEN

Cycle 61 R = -1.999999999999999999999999999999999999997215600587600481445798530176882707661 M = 1.91x10^{71} S = 100x76 D = 1500 AGDP ZEN

Cycle 62 R = -1.99999999999999999999999999999999999999930390014690012036144963254422067692 96 M = 3.06x10^{72} S = 100x76 D = 1500 AGDP ZEN

Cycle 63 R = -1.99999999999999999999999999999999999999982597503672503009036240813605516923331 M = 4.88x10^{73} S = 100x76 D = 1500 AGDP ZEN UC

Cycle 64 R = -1.9999999999999999999999999999999999999999564937591812575225906020340137923083 86 M = 9.78x10^{74} S = 100x76 D = 1500 AGDP ZEN UC

Cycle 65 R = -1.99999999999999999999999999999999999999998912343979531438064765050850344807710022 M = 1.56x10^{76} S = 100x76 D = 1500 AGD ZEN UC

Cycle 66 R = -1.99999999999999999999999999999999999999999728085994882859516191262712586201927 5289 M = 2.50x10^{77} S = 100x76 D = 1500 AGDP ZEN UC

Cycle 67 R = -1.999999999999999999999999999999999999999999320214987207148790478156781465504818 83720 M = 4.00x10^{78} S = 100x76 D = 1500 AGDP ZEN UC

Cycle 68 R = -1.999999999999999999999999999999999999999999830053746801787197619539195366376204 710240 M = 6.41x10^{79} S = 100x76 D = 1500 AGDP ZEN UC

Cycle 69 R = -1.99999999999999999999999999999999999999999995751343670044679940488479884159405117 76196 M = 1.03x10^{80} S = 100x76 D = 1500 AGDP ZEN UC

Cycle 70 R = -1.99999999999999999999999999999999999999999999999893783591751116998512211997103985127944 08699 M = 1.64x10^{82} S = 100x76 D = 1500 AGDP ZEN UC

Cycle 71 R = -1.9999999999999999999999999999999999999999999999999734458979377792496280529992759962819860 24147 M = 2.62x10^{83} S = 100x76 D = 1500 AR VAX

Cycle 71 R = -1.9999999999999999999999999999999999999999973445897937779249628052999275996281986024147 M = 2.62×10^{83} S = 100x76 D = 1500 AR VAX

Cycle 72 R = -1.99999999999999999999999999999999999999999336147448444481240701324981899907049 65061889 M = 4.20×10^{84} S = 100x76 D = 1500 AR VAX

Cycle 73 R = -1.99999999999999999999999999999999999999999834036862111120310175331245474976762 412655687 M = 6.72×10^{85} S = 100x76 D = 1500 AR VAX

Cycle 74 R = -1.99999999999999999999999999999999999999999958509215527780077543832811368744190 6031639829 M = 1.08×10^{87} S = 100x76 D = 1500 AR VAX

Cycle 75 R = -1.99999999999999999999999999999999999999999989627303881945019385958202842186047 65079099961 M = 1.72×10^{88} S = 100x76 D = 1500 AR VAX

Cycle 76 R = -1.999999999999999999999999999999999999999999997406825970486254846489550710546511 912697750149 M = 2.75×10^{89} S = 100x76 D = 1500 AR VAX

Cycle 77 R = -1.999999999999999999999999999999999999999999999351706492621563711622387677636627 9781744375528 M = 4.40×10^{90} S = 100x76 D = 1500 AR VAX

Cycle 78 R = -1.999999999999999999999999999999999999999999999837926623155390927905596919409156 99454360938919 M = 7.04×10^{91} S = 100x76 D = 1500 AR VAX

Cycle 79 R = -1.999999999999999999999999999999999999999999999959481655788847731976399229852289 2486359023473600 M = 1.13×10^{93} S = 100x76 D = 1500 AR VAX

Cycle 80 R = -1.999999999999999999999999999999999999999999999989870413947211932994099807463072 31215897558684395 M = 1.80×10^{94} S = 100x76 D = 1500 AR VAX

Cycle 81 R = -1.999999999999999999999999999999999999999999999997467603486802983248524951865768 0780397438967112396 M = 2.89×10^{95} S = 100x76 D = 2000 AR VAX

Cycle 82 R = -1.999999999999999999999999999999999999999999999999366900871700745812131237966442 01950993597417782575 M = 4.62×10^{96} S = 100x76 D = 2000 AR VAX

Cycle 83 R = -1.999999999999999999999999999999999999999999999999841725217925186453032809491610 504877483993544457441 M = 9.24×10^{97} S = 100x76 D = 2500 AGDP ZEN

Cycle 84 R = -1.999999999999999999999999999999999999999999999999960431304481296613258202372902 62621937099838611442374 M = 1.18×10^{99} S = 100x76 D = 2000 AR VAX

Cycle 85 R = -1.999999999999999999999999999999999999999999999999990107826120324153314550593225 656554842749596528609948 M = 1.89×10^{100} S = 200x151 D = 2000 AR VAX

Cycle 95 R = -1.999999999999999999999999999999999999999999999999999999990566087837528374971914 8571259084270884986844335663868127676 M = 2.08×10^{112} S = 100x76 D = 3000 AGDP ZEN UC

Cycle 100 R = -1.999999999999999999999999999999999999999999999999999999999990787195153836303683 51060266201994832861199651715467462319416670 M = 2.18×10^{118} S = 100x76 D = 3000 AGDP ZEN UC

Cycle 105 R = -1.999999999999999999999999999999999999999999999999999999999999991003120267418265 3159283229121288557896601528487838619367144106110 M = 2.29×10^{124} S = 100x76 D = 3000 AGDP ZEN UC

Cycle 110 R = -1.99999999999999999999999999999999999999999999999999999999999999999991213984636150
649722586252843875835732089993016390490172572667972491 M = 2.40×10^{130} S = 100x76 D = 3000
AGDP ZEN UC

Cycle 115 R = -1.99999999999999999999999999999999999999999999999999999999999999999999991419906871
2408688697131375428474958321191338050688380591529960682462 60 M = 2.51×10^{136} S = 100x76
D = 3000 AGDP ZEN UC

Cycle 120 R = -1.99999999999999999999999999999999999999999999999999999999999999999999999991621002
80394616100557923588168700764855384160651253716714159772289809536 0 M = 2.64×10^{142} S = 100x76
D = 3000 AGDP ZEN + ZEN UC

Cycle 125 R = -1.99999999999999999999999999999999999999999999999999999999999999999999999999991817
3855507286728570109725407099684067908609438598995772867165262676726 0 M = 2.67×10^{148} S = 100x76
D = 3000 AGDP ZEN UC

Cycle 150 R = -1.99999999999999999999999999999999999999999999999999999999999999999999999999999999
9999999999927323784294307821136724468231360138108872889829291881878307859621447951453738355
73581117679592 M = 1.78×10^{179} S = 100x76 D = 10000 AGDP ZEN UC

Cycle 175 R = -1.99999999999999999999999999999999999999999999999999999999999999999999999999999999
99999999999999999999999999354505535847328988700905833931643745672283161856447381040061547 18
781786042692477188441548689870613445768523494 M = 4.44×10^{208} S = 100x76 D = 10000
PR ZEN + VAX

Cycle 200 R = -1.999999999999999999999999999999999999999999999999999999999999999999999999999999999
9999999999999999999999999999999999999999426685746903701524868873414969921982795393910464951 5
64378620882667385903299117419504340886237075855598460105829200658465344388 M = 5.68×10^{238}
S = 100x76 D = 10000 RP DEC SYS5000 + ZEN

Cycle 250 R = -1.999999999999999999999999999999999999999999999999999999999999999999999999999999999
99999999999999999999999999999999999999999999999999999999999999999999995477347993263457146128
49868772624150857945748314942293641602039247450498736039839970632632438592848265993465823922
757562994579389408988813808014051893359742 M = 9.04×10^{298} S = 100x76 D = 10000
RP DEC SYS5000 + ZEN

Cycle 300 R = -1.999999999999999999999999999999999999999999999999999999999999999999999999999999999
99999999999999999999999999999999999999999999999999999999999999999999999999999999999999999969
42265581617082564244406089532095341856110462695594154906282867945161787009721229863548534802
541940266148691731543402237271898408295765449289538680351633 M = 1.45×10^{359} D = 10000
AR DEC SYS5000

INDEX

Under the listing of Midgets, Cycles an underlined page number indicates a picture of that midget exists on the underlined page.

INDEX